DESIGN AND DEVELOPMENT OF TWO NOVEL CONSTRUCTED WETLANDS: THE DUPLEX-CONSTRUCTED WETLAND AND THE CONSTRUCTED WETROOF

MARIBEL ZAPATER PEREYRA

Thesis committee

Promotor

Prof. Dr P.N.L. Lens
Professor of Environmental Biotechnology
UNESCO-IHE Institute for Water Education
Delft, the Netherlands

Co-promotor

Dr J.J.A. van Bruggen
Senior lecturer in Microbiology
UNESCO-IHE Institute for Water Education
Delft, the Netherlands

Other members

Prof. Dr A. van der Wal, Wageningen University
Prof. Ing. J. Vymazal, Czech University of Life Sciences, Prague, Czech Republic
Dr N. Fonder, ENvironnement Économie Ecologie, Namur, Belgium
Prof. Dr J.T.A. Verhoeven, Utrecht University

This research was conducted under the auspices of the SENSE Research School for Socio-Economic and Natural Sciences of the Environment

DESIGN AND DEVELOPMENT OF TWO NOVEL CONSTRUCTED WETLANDS: THE DUPLEX-CONSTRUCTED WETLAND AND THE CONSTRUCTED WETROOF

Thesis
submitted in fulfilment of the requirements of
the Academic Board of Wageningen University and
the Academic Board of the UNESCO-IHE Institute for Water Education
for the degree of doctor
to be defended in public
on Friday, 30 October 2015 at 4 p.m.
in Delft, the Netherlands

by

Maribel Zapater Pereyra
Born in Piura, Peru

CRC Press/Balkema is an imprint of the Taylor & Francis Group, an informa business

© 2015, Maribel Zapater Pereyra

Published by:
CRC Press/Balkema
PO Box 11320, 2301 EH Leiden, the Netherlands
e-mail: Pub.NL@taylorandfrancis.com
www.crcpress.com – www.taylorandfrancis.com

ISBN 978-1-138-02930-9 (Taylor & Francis Group)
ISBN 978-94-6257-496-0 (Wageningen University)

To mi chico.

ACKNOWLEDGEMENTS

I would like to start addressing my sponsor that financially supported this study: UNESCO–IHE Partnership Research Fund project No. 32019417 NATSYS (Natural Systems for Wastewater Treatment and Reuse: Technology Adaptation and Implementation in Developing Countries), The Netherlands.

My sincere gratitude goes to my promotor Prof. Dr. P.N.L. Lens for his guidance, feedback and training along all these years. As well to my supervisor and friend Dr. J.J.A. van Bruggen. This dissertation could not have been finished without your help. Thanks Hans!

Many thanks to Prof. Dr. ir. Diederik Rousseau for his supervision (and the beers we shared) at the beginning of my PhD. Also to my external supervisors, Dr. Carlos Arias and Scott Wallace, I would like to show my deepest gratitude. Both, despite the distance, were always there to help me when I contacted them. Thanks Scott for helping me with the "mysteries" of aeration.

Over the past years I have received excellent support by all my MSc. students via the data gathered, the ideas developed and shared, the funny lab working moments, the long but nice meetings and the paper writing. In chronological order of supervision I would like to thank: Emanuela Malloci, Stevo Lavrnić, Huma Ilyas, Emma Kyomukama, Violet Namakula and Peter Pintér. Thanks for all the shared time and for teaching me to improve my supervision skills. I have learnt so much from you. Many thanks as well to my co-author Elisée Gashugi for his dedication and support during the paper writing.

None of my results could have been possible without the help of all lab members: thanks Fred, Peter, Don, Berend, Ferdi, Frank and Lyzette. Special thank goes to Mr. Berthold Verkleij that made all my research possible since he helped me to bring almost 360 L of wastewater every week. Thanks a lot as well to Jolanda Boots and Sylvia van Opdorp-Stijlen for your big help with different issues along my PhD. Kirstin, thanks for the quick English-editing you helped me with. I would like to express my deep appreciation and gratitude to Eldon Raj for reading some of my papers during our paper-writing meetings. What a help Eldon, thanks a lot!

Frank van Dien was a great piece in my PhD. He not only believed in me but constantly proved it since, for instance, he introduced me to the constructed wetroof project and gave me the opportunity to work with him in his company ECOFYT as part of my PhD training. It has been a pleasure to work with you in different projects and thanks for allowing me to work with real scale constructed wetlands and for the partial funding that made me reach Australia. I will never forget that!

Special thanks to the people from (or related to) van Helvoirt Groenprojecten that made the constructed wetroof research possible: Remco Valk, Bert van Helvoirt, Joris Voeten, Harald

Waijers, Frans van Abeelen and Annemiek Verbunt. Without this great team of people, none of the constructed wetroof findings would have become real. Thanks for the permanent help, smiles, hopes, encouraging words and transportation funding.

I met so many friends along this journey that turned into an important piece in this PhD puzzle: Male, Caro, Juliette, Jorge, Vivi, Vane, Lyda, Lui, Isa, Mijail, Vero, Fer, Erika, Jessi, Angélica, Pato, Eva, Rossana, Mauri, Alex, Bro, Raquel, Jona, Connie, Volker and Theresita! Thank you so much for the moments we have shared that gave me the strength to continue.

My family, the hidden motor of this journey. Mom thanks for being the best mom ever, enjoying with me my happy and successful moments and hating all what made me suffer. Te quiero! To my grandparents, Mamameli and Papacarlos, my beloved "viejitos": thanks for being the coolest grandparents ever that allow us to skype almost every day to share our overseas-anecdotes. Mire, Kiara, Coco, mi Vale hermosa: maybe you do not know but our moments together, skyping or "whatsapping" gave me always the fresh mind I needed to continue with this... Gracias!

And finally, to my love, to whom I dedicate my thesis: mi chico! There are no words to describe all what you have done for me along this journey. From the bottom of my heart, VIELEN DANK! Thanks for being such a good person and partner. This PhD was certainly difficult, but without any doubt you made it easier! This is our PhD Dr. Jojo! Te amo!

SUMMARY

Constructed wetlands (CWs) are man-made technologies, which mimic natural systems and their processes, to treat wastewater flowing through a filter media and plants. CWs are among the few natural treatment systems that can guarantee an efficient wastewater treatment and an appealing green space at the same time. Ergo, they are of worldwide importance since they can facilitate water reuse while freeing fresh water sources for potable use and, in the case of many urban areas, they can restore the lost green areas caused by extensive construction. However, CWs require large spaces for their construction which usually are not available in urban areas.

This downside of CWs has been the motivation to develop wastewater treatment options with less space requirement. In this thesis, two options were designed and studied: the Duplex-CW and the constructed wetroof (CWR). The Duplex-CW is a hybrid CW composed of a vertical flow (VF) CW on top of a horizontal flow (HF) filter (HFF). The stacked arrangement is the key for reducing the CW footprint, since the two systems occupy the same space by having a vertical arrangement instead of the common horizontal arrangement. The CWR is a shallow HF CW placed on the roof of a building, thus it does not occupy any land. Along this document, several modifications and improvements have been tested in order to select the most appropriate Duplex-CW and CWR design for the treatment of domestic wastewater.

In this study, the development of a compact CW (Chapter 3-6) demanded the evaluation of different design aspects such as arrangement modes (i.e. conventional and stacked for a single and hybrid system, respectively), configurations of the stacked arrangement (i.e. fill and drain -Fill&D-, stagnant batch and free drain), loading rates (i.e. hydraulic -HLR- and organic -OLR-), intensification (i.e. artificial aeration and recirculation), carbon source as electron donor for denitrification and post-treatment for phosphorus removal. Altogether, the stacked Fill&D Duplex-CW outperformed all other tested CWs due to the combination of the intermittent feeding, resting periods and a 1 d hydraulic retention time (HRT) in the VF CW with the saturated conditions and HRT of 3-4 d in the HFF. Artificial aeration, recirculation and the addition of a carbon source as an electron donor for denitrification were not necessary in the design with current operational conditions. The effluent quality achieved (41-69 mg L^{-1} for chemical oxygen demand (COD), 8-16 mg L^{-1} for total suspended solids (TSS), 6-24 mg L^{-1} for NH_4^+-N, 13-24 mg L^{-1} for total nitrogen (TN) and 2-5 mg L^{-1} for total phosphorus (TP)) met European discharge guidelines treating OLRs up to 37 g COD m^{-2} d^{-1} except for TP. Therefore, the recommended design requires an area of 2.6 m^2 PE^{-1} (population equivalent) and a post-treatment filter for phosphorus removal.

Efforts to develop the CWR (Chapter 7-8) included testing the mass, hydraulic and stability properties of different materials, the treatment capacity of each material and the effect of sun and rain on the treatment capacity and hydrology of the system. Overall, the CWR design consisted of a shallow (9 cm) filter bed (i.e. sand, light expanded clay aggregates -LECA- and polylactic acid beads) holding in together by means of the grass roots and stabilization plates.

The intrinsic aeration guaranteed an efficient removal of organic matter (79-82% for COD) and nutrients (> 99% for NH_4^+-N). The phosphorus adsorption capacity of LECA explained the high TP removal (86-97%). For TN it was postulated that aerobic denitrifiers could have be present in the CWR. The CWR's hydrology was affected by the presence of sun and rain, but it did not deteriorate the effluent quality.

In synthesis, this thesis demonstrates that the Fill&D Duplex-CW and the CWR are efficient domestic wastewater treatment technologies with a lower space requirement (< 2.6 m^2 PE^{-1}) than common European design (5 m^2 PE^{-1}). However, the Duplex-CW area requirement is still higher than many CW designs and therefore further improvements should be directed at increasing the loading rate and the VF CW HRT to decrease the area requirements. The CWR is the foremost option to save land areas since it requires 0 m^2 of land PE^{-1}.

SAMENVATTING

Kunstmatige rietvelden zijn geconstrueerde systemen om afvalwater te zuiveren door middel van een filtermedium met planten en zij bootsen hiermee natuurlijke systemen (moerassen) en hun processen na. Rietvelden behoren tot de weinige natuurlijke systemen die een efficiënte afvalwaterzuivering met een aantrekkelijke groene ruimte combineren. Ze zijn van wereldwijd belang omdat zij hergebruik van water mogelijk maken. Ook bieden ze de mogelijkheid in veel stedelijke gebieden om groene ruimtes te creëren. Rietvelden vereisen echter zelf veel ruimte, wat vaak niet aanwezig is in die stedelijke gebieden.

De keerzijde van rietvelden (het ruimtebeslag) is de stimulans geweest om afvalwaterzuiveringssystemen te ontwikkelen die minder ruimte nodig hebben. In dit proefschrift zijn twee mogelijke systemen ontwikkeld en getest: het gecombineerde rietveld (Duplex rietveld) en een zogenaamd geconstrueerd Nat Dak Systeem (NDS). Het Duplex systeem is een hybride rietveld samengesteld uit een vertikaal doorstroomd rietveld boven op een horizontaal doorstroomde filter. Dit gecombineerde systeem is de sleutel voor het verminderen van de voetprint van het rietveld, aangezien de twee systemen dezelfde ruimte in beslag nemen als het algemeen gebruikte rietveld. Het NDS is een ondiep horizontaal doorstroomd systeem, beplant met gras, en aangelegd op een dak. Hierdoor neemt dit systeem netto totaal geen ruimte in beslag.

In deze studie vereiste de ontwikkeling van een compact rietveld (Hoofdstukken 3 tot 6) de bestudering van verschillende ontwerp aspecten, zoals opstellingen (bijvoorbeeld het conventionele en gestapelde systeem, voor respectievelijk het enkelvoudige en hybride systeem), configuraties van het Duplex systeem (bijvoorbeeld vul en leegloop, blijvend gevuld en vrije leegloop), belasting (hydraulisch en organisch), intensivering (beluchting en recirculatie), koolstof bron als electron donor voor de denitrificatie en nazuivering voor de verwijdering van fosfaten. Samengevat was de zuivering door het Duplex gestapelde vul en leegloop systeem van alle geteste systemen het beste door een combinatie van a) de toevoer van afvalwater met tussenpozen en b) een hydraulische verblijfstijd van 1 dag voor het verticaal doorstroomd rietveld en c) een verblijfstijd van 3 tot 4 dagen voor het horizontaal doorstroomd filter. Beluchting, recirculatie en de toevoeging van een koolstofbron als electron donor voor denitrificatie was niet noodzakelijk in het ontwerp onder de beschreven operationele condities. De verkregen kwaliteit van het effluent (41-69 mg/L voor het CZV (chemisch zuurstof verbruik), 8 - 16 mg/l voor de totaal opgeloste stoffen, 6 - 24 mg/l voor NH_4^+ stikstof, 13 - 24 mg/l voor totaal stikstof en 2 - 5 mg/l voor totaal fosfaat voldeed aan de Europese standaard voor een organische belasting tot 37 gram chemisch zuurstof verbruik per m^2 per dag, uitgezonderd voor totaal fosfaat. Om deze reden wordt voor het definitieve ontwerp met een oppervak van 2.6 m^2 per populatie equivalent gerekend en is een nazuiveringsfilter voor de verwijdering van fosfaat noodzakelijk.

Onderzoek naar het beste ontwerp voor een geconstrueerd Nat Dak Systeem (Hoofdstuk 7 en 8) omvatte het gewicht, hydrologie en eigenschappen met betrekking tot de stabiliteit van verschillende materialen, de zuiveringscapaciteit van elk materiaal en de invloed van zon en regen op de zuiveringscapaciteit en hydrologie van het systeem. Het NDS bestond uiteindelijk uit een ondiepe (9 cm) filter laag (zand, lichte geëxpandeerde klei korrels - LECA- en polylactaat bolletjes) samengehouden door graswortels en stabilisatieplaten. De intrinsieke beluchting garandeerde een efficiënte verwijdering van organisch materiaal (79 - 82 % voor het CZV) en nutriënten (> 99 % voor NH_4^+-N). De adsorptie capaciteit voor fosfaat van LECA verklaarde het hoge verwijderingpercentage (86 - 97 %) voor TP. Aërobe denitrificerende bacteriën waren waarschijnlijk aanwezig en verantwoordelijk voor de verwijdering van totaal stikstof (TN). De hydrologie van het NDS wordt beïnvloed door zonneschijn en regenval, maar de effluent kwaliteit werd er niet door beïnvloed.

Dit proefschrift illustreert dat het vul en leegloop Duplex rietveld en het NDS efficiënte afvalwaterzuiveringstechnieken zijn voor huishoudelijk afvalwater met minder ruimtebeslag (<2.6 m^2 per PE) dan de algemene Europese ontwerp richtlijn (5 m^2 per PE). Het ruimtebeslag voor het Duplex systeem is echter hoger dan veel rietveld ontwerpen en daarom moeten verdere verbeteringen gericht zijn op het verhogen van de belasting en de hydraulische verblijfstijd om het ruimtebeslag te verkleinen. Het NDS is de voornaamste optie om ruimte te besparen aangezien het uitsluitend nutteloze ruimte in beslag neemt, zodat men kan stellen dat het oppervlakte beslag 0 m^2 per PE wordt.

CONTENTS

CHAPTER 4.   EFFECT OF AERATION ON POLLUTANTS REMOVAL, BIOFILM ACTIVITY AND PROTOZOAN ABUNDANCE IN CONVENTIONAL AND HYBRID HORIZONTAL SUBSURFACE-FLOW CONSTRUCTED WETLANDS

CHAPTER 7.   MATERIAL SELECTION FOR A CONSTRUCTED WETROOF RECEIVING PRE-TREATED HIGH STRENGTH DOMESTIC WASTEWATER .. 105

CHAPTER 8.   CONSTRUCTED WETROOFS: A NOVEL APPROACH FOR THE TREATMENT AND REUSE OF DOMESTIC WASTEWATER AT HOUSEHOLD LEVEL   115

xviii

CHAPTER 1.

GENERAL INTRODUCTION

1.1. NEED OF NATURAL WASTEWATER TREATMENT SYSTEMS

Statistics figures of the raising population number in the world are alarming. It is expected that in 2050 the amount of inhabitants in the world will reach approximately 9 billion, from the 6.6 billion existing to date. In other words, the population growth speed can be estimated like adding, on average, the population of Italy each year to the world (United Nations, 2004, 2005). Among the variety of impacts that such growth can bring, three main direct consequences exist in the water sector: (i) increment of fresh water demand and decrease of fresh water sources, (ii) larger wastewater production and necessity of more sanitation technologies and, (iii) reduction of (green) area due to massive construction of houses, offices or industries to satisfy the needs of the increasing number of people.

Incredible efforts are being done to guarantee the access to water and sanitation to everybody. The Millennium Development Goal Target 7c established that for 2015 the amount of people without access to fresh water and to sanitation should be halved. The access to drinking water has already been fulfilled in 2010, but still 780 million people do not have access to an improved drinking water source. The sanitation requirement appears to fail by 2015 (reaching only 67% of the task) as nowadays, still 2.5 billion people lack improved sanitation (UNICEF and WHO, 2012). This situation can aggravate further due to the current worldwide situation that the majority of the world's population is living in urban areas (WHO and UN-HABITAT, 2010). Governments are unable to control the rapid and usually informal urban growth. Therefore, formation of cities with poorly designed water-, sanitation- and transport-systems occurs (WHO and UN-HABITAT, 2010). Overpopulated and inappropriately constructed cities can spread infectious diseases easily (malaria, diarrhea and worm infections) and can decrease the amount of (green) spaces within the city leading into an increment of grey-architecture (concrete).

A clever approach to minimize those negative impacts in the (city) environment is by encouraging the natural treatment of the wastewater. Natural wastewater treatment systems (NWTS) are technologies used for the treatment of (many types of) wastewater in an environmental friendly way, e.g. exploiting the benefits of natural processes without using external chemical sources. Such technologies can be treatment ponds, constructed wetlands (CWs), living machines and soil-aquifer treatment. The systems can provide the treatment for wastewater, meet the sanitation requirements and supply water resources for different population requirements.

The added value of some NWTS, namely CWs, is the aesthetic benefits that the use of macrophytes provides to the surroundings; offering "green" to the city concrete. Thus, the integration of planted NWTS within the urban environment could promote sanitation, increase the amount of green spaces and enhance the quality of living; topics that are part of the Water Sensitive Urban Design (WSUD) concept. WSUD was a recently developed (in the 90's) initiative in Australia that involves social and environmental sciences *"for a holistic management of the urban water cycle* (potable, sewerage and storm water) *and its integration into urban design"* (Wong, 2007).

Additionally, NWTS have the great advantage of not using (a lot of) electricity and therefore being environmental friendly and economic, but, at the same time, have the disadvantage of occupying more area or volumes to enable an efficient wastewater treatment than other wastewater treatment systems, such as the activated sludge system (Mara, 2006). This variation in space is due to the fact that without external sources for intensifying the treatment (e.g. electricity and chemical addition) bigger systems are needed to compensate for those requirements. But in the long-run, they are most cost-effective systems, while other treatment systems may not demand a high investment for land requirement but instead require permanent investment for their operation and maintenance.

A brief case study that highlights the need of CWs and WSUD in a megacity is presented below.

Megacity of Lima, Peru: Green areas irrigation - case study
Lima, the capital of Peru, is a rapidly growing metropolis of currently 9 million inhabitants (14 times more than in 1940) located at the desert coast of the country (Ioris, 2012a,b). The desert of Lima suffers from water stress due to the fact that almost no precipitation occurs (< 15 mm yr^{-1}) and that the three rivers that provide the majority of the water to the city (Rímac, Chillón and Lurín) are polluted and (almost) dry during nearly all the year (0-10 m^3 sec^{-1} from May to December) (Fernández-Maldonado, 2008). Furthermore, global warming has caused the loss of millions of cubic meters of water in the Andean glaciers resulting in not only a concern about water availability for the three rivers, but also conflicts with the energy sector (Vergara et al., 2010) since approximately 60% of electricity in Peru is currently generated using hydropower (MEM, 2012).

The water delivery infrastructure in the metropolitan area currently serves only 84.3% and 80.1% of inhabitants with drinking water and sanitation, respectively due to the rapidly settlement establishments (Ioris, 2012a). The remaining people, living in the most peripheral area that are difficult to reach, get expensive and limited water delivered by trucks (up to 9 times more than water delivered via the city's pipe distribution network and of low quality) (Fernández-Maldonado, 2008; Ioris, 2012b). Furthermore, in Lima only 2.4 m^2 (in average) of green space per person is available (Moscoso, 2011) and despite the water stress, fresh water is the main water source for irrigation.

The background of Lima requires urgent solutions that can alleviate the current situation. An appropriate option, within the frame of WSUD, is the application of CWs: green technologies easily integrated within the surroundings that can provide green and recreational areas in the desert of Lima, supply the required sanitation needed (mainly in the peripheral areas) and give the opportunity for water reuse (for more green spaces) (Eisenberg et al., 2014).

1.2. CONSTRUCTED WETLANDS: GENERAL OVERVIEW

CWs are low-cost and environmentally friendly NWTS that are based on the concept of natural wetlands. The interaction of plants, microorganisms and soil results in natural processes (biological, chemical and physical) are able to remove pollutants from wastewater (Fig. 1.1). CWs do not require a skilled operator to be maintained, can be placed onsite and promote sustainable rural sanitation and the decentralization of wastewater treatment (Kadlec and Wallace, 2009).

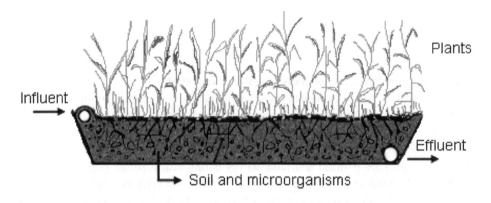

Figure 1.1 Schematic representation of a subsurface flow constructed wetland.

The first type of CW used was the free-water surface flow (FWS) or surface flow system. In FWS systems the wastewater flows above the soil substrate limiting its contact with the CW substrate and exposing the water to the environment, thus increasing the potential problems of insects, odors, ecological risks and human exposure. Later, the sub-surface water flow (SSF) CW was developed, with the water flowing horizontally through the system (Kadlec and Wallace, 2009). The horizontal flow (HF) CW was designed expecting enough oxygen (from the roots and rhizomes of the macrophytes) for aerobic degradation and nitrification. However, the amount of oxygen released by the plants was overestimated and instead, an efficient denitrification and anaerobic organic matter degradation was achieved as the majority of the saturated bed was anaerobic (Vymazal, 2005). Later, the interest in producing systems with well nitrified effluents substituted the traditional CWs (FWS and HF CW) for a vertical flow (VF) CW (Cooper, 1999). The VF CW provided a greater supply of oxygen that increased the nitrification rate and the performance of aerobic biological oxidation that resulted in a good removal of biochemical and chemical oxygen demand (BOD_5 and COD, respectively).

Despite all the efforts along the time in producing the best effluent quality, some limitations of all types of CWs remained inherited in the systems such as bed clogging, yet limiting oxygen transfer, poor pathogen and nutrient removal and a relatively high land area required.

The *space required* (from now on called, *footprint*) of CWs becomes a major disadvantage in megacities where not only the land price is a limitation but also the land availability. Availability in the sense of land scarcity and/or the preference to use the land for other purposes (e.g. open-, environmental- and economic-spaces). In rural areas, this situation is not problematic as they are usually surrounded by empty areas. Nevertheless, geographically it can be a problem as maybe the topography of the area does not allow for the construction of wastewater treatment systems with a large footprint.

As a rule of thumb, the area needed by a conventional HF CW is approximately 5 m^2 per population equivalent (PE) while a VF CW requires a lower area demand (1-3 m^2 PE^{-1}; Vymazal, 2011) due to its higher bed oxygenation. In any case, these values are too large when compared for instance with activated sludge systems that require only 0.2-0.4 m^2 PE^{-1} (Mburu et al., 2013). Thus, only innovative CWs are strongly required to meet the treatment requirements within the space that is available. This research studies solutions to reduce the land requirements by the CWs.

1.3. SCOPE AND OBJECTIVES OF THIS THESIS

Population growth is leading to the formation of crowded cities, thus reducing space availability in a city for any purpose and demanding additional drinking water and sanitation technologies. CWs are an alternative but unfortunately they occupy more space than other wastewater treatment technologies. Hence, the overall aim of this research is *"To develop new CW designs capable of overcoming the limitation of large system footprint during the treatment of domestic wastewater reaching similar or better effluent quality than that in conventional CWs"*.

This research will design, develop and investigate the performance of two systems for domestic wastewater treatment: (i) a Constructed Wetroof (CWR), a modification of Green Roofs (GR) with the benefit of CWs and (ii) a Duplex-CW, a hybrid CW in a stack up design.

A CWR is a CW placed on a roof of a building with the main purpose of wastewater treatment. The design of a CWR is highly dependent on the weight requirements of a building. Thus, it inclines towards a very shallow design filled with light-weight materials. Using a CWR saves 100% of the land area that a conventional CW would have needed, as it is placed on a commonly unused spaced: roof of a building.

A Duplex-CW is a hybrid CW that, instead of placing the different systems (or compartments) next to each other on the ground, organizes them vertically (one on top of the other). Thus, providing similar treatment as a hybrid CW without increasing the land usage.

1.4. THESIS OUTLINE/ STRUCTURE

This thesis is divided in 9 chapters. **Chapter 1**, the Introduction, consists of a brief explanation of the background of this research. The concept of NWTS, such as CWs, is introduced. **Chapter 2** is a literature overview of all the efforts conducted over the years in improving the performance of conventional CWs by including special add-ons in their design. **Chapter 3** explains the study of different materials for phosphorus removal from wastewater. The results of this chapter will be used as part of the design of the system studied in chapters 5 and 6. **Chapter 4** comprises the information obtained from the aeration studies done to three HF CWs for testing their performance, biofilm activity and protozoan community in order to draw conclusions about the best system performance for footprint reduction. **Chapters 5's** main focus is the development of a CW, called Duplex-CW, with potential benefits in improving pollutant removal while decreasing the footprint needed. **Chapter 6** contains information about configurations of Duplex-CWs that differ in their mode of operation and presents the recommended Duplex-CW design. **Chapter 7** gathers the information collected during several trials that were done to select the most appropriate material used in the CWR. **Chapter 8** explains the challenges of the CWR and the experiments that followed in order to ascertain the strengths and weakness of the system in order to improve its design. **Chapter 9** contains an overall discussion integrating all previous chapters and the conclusions and recommendations, presenting the final outcome of this PhD research.

CHAPTER 2.
LITERATURE REVIEW

2.1. INTRODUCTION

There are many wastewater treatment configurations and designs, all possess certain drawbacks depending on the type of system, purpose of it, weather conditions where it is used and type of wastewater it treats. The constructed wetland (CW) technology is not an exception and one of the main existing drawbacks will be tackled in this chapter: the large area demand. In some situations, this drawback can be the reason for discarding the use of CWs; for instance:

▪ In big cities and megacities, where space is scarce and the size of the wetland is dictated by the existing streets and buildings (Kadlec and Wallace, 2009).

▪ In situations where local authorities demand to the property owner a certain volume and/or quality of wastewater to be discharged, but the property lacks the space for a large treatment system.

▪ In situations where the land owner prefers to use part of the land for other purposes rather than for a wastewater treatment system only.

▪ In mountain regions. Foladori et al. (2012) explained that, for 1200 PE (population equivalent) using a design of 4 m^2 PE^{-1}, 4800 m^2 of contiguous land surface is needed and this is rarely found in a mountain region due to slope characteristics and the importance of conserving wildness.

For those situations, compact CWs are considered a relevant solution that can contribute to sanitation and, at the same time, can create more green spaces. Thus, this literature review will present different improvements done to CWs in order to improve their performance. The general idea is to maximize the treatment performance of each square meter of the CW, consequently requiring less area for the treatment of the same amount of wastewater. In other words, a lower m^2 PE^{-1} would be required.

A conventional CW design requires large land areas to increase the chances for having a good effluent quality. As a rule of thumb, the area needed by a horizontal flow (HF) CW is approximately 5 m^2 per PE and a vertical flow (VF) CW requires between 1 and 3 m^2 PE^{-1} (Vymazal, 2011). These values are too large when compared for instance with activated sludge systems that require 0.2-0.4 m^2 PE^{-1} (Mburu et al., 2013) or even less (0.062-0.17 m^2 PE^{-1}; Cooper, 2005). For a comparison of the area requirement by different technologies, see Table 2.1.

Table 2.1 Area requirements per population equivalent (m^2 PE^{-1}) by different treatment technologies. Taken from Cooper (2005).

Level of treatment	Constructed wetland	Biofilter	Activated sludge	Biological aerated filter	Biological fluidized beds
BOD removal only	5 (HF) 1 (VF)	0.40	0.062	0.027	0.013
Nitrification + BOD removal	1-2 (VF)	1.14	0.17	0.081	0.039

PE, population equivalent; BOD, biochemical oxygen demand; HF, horizontal flow; VF, vertical flow.

The area can even be further enlarged if different CW stages are necessary, for example one stage that can provide aerobic conditions and another with an anoxic environment. As land availability and cost may limit the use of such systems, new approaches are necessary for enhancing the CW performance without compromising the required area. According to Wu et al. (2014), research groups and published literature trying to improve the treatment performance with new possible solutions has drastically increased. Most, but not all of these solutions require an intensification (use of energy) usually characterized by a high cost.

The approaches can be summarized into three categories: (i) to *boost the treatment efficiency* of the system, (ii) to *stack up* extra treatment stages (increase the CW depth or height) without increasing the system footprint, and (iii) to place the CW at *unused spaces*. Examples of them are found below. For the majority of them, the area required per PE is obtained directly in the manuscript. When this information was not available it was calculated. The footprint taken by a CW depends on many factors such as the level of treatment (i.e. primary, secondary and tertiary), type of wastewater to be treated (e.g. domestic, industrial, agricultural, dairy, acid mine drainage and landfill leachate), climate and the methodology used for its calculation (Cooper, 2005; Kadlec and Wallace, 2009). In this chapter, for simplicity and due to the limited data, calculations of PE followed Equation 2.1 unless otherwise stated.

1 PE = 60 g BOD_5 d^{-1} Eq. 2.1

Where,

PE, population equivalent

BOD_5, 5-day biochemical oxygen demand

2.2. BOOSTING THE TREATMENT EFFICIENCY

Usually, to enhance the system efficiency it is proposed to design it with a considerable amount of (additional) oxygen or to increase the contact time of the water and the filter media. Enough oxygen within the system gives the microbiota the conditions to complete biodegradation, decreases the possibilities for clogging and, at the same time, enhances the system efficiency. A larger contact time (e.g. achieved by recirculating the wastewater) increases the chances of the pollutants to be treated by filter media and related biofilm. Therefore, some of these strategies are reviewed in this section.

2.2.1. Recirculation

Sun et al. (2003) studied a three-stage VF CW with and without recirculation for diluted pig slurry from a farm (Table 2.2, Fig. 2.1). In the recirculating system the wastewater enters the first stage, flows throughout the bed, is recycled back to the top of the bed, flows once again throughout the bed and then leaves the system to the following stages, where the same cycle will be repeated. Without recirculation, the wastewater was only flowing from one stage to the other. The design required 1.5 m^2 PE^{-1} (conventional CW) and 2.2 m^2 PE^{-1} (with recirculation), however with recirculation, a better performance was achieved (Table 2.2). The conclusion achieved was that recirculation increased the systems efficiency because it increases the contact time of the pollutants with the bed media and increases the bed oxygenation. Thus, a better water quality can be achieved without increasing the CW area, since the water passes through the filter more than once. Similarly, plenty of studies with recirculation in different CW setups have made similar conclusions; see Section 2.1.3 for the recirculating tidal CW (Zhao et al., 2004a; Sun et al., 2005) and Section 2.2 for the recirculating stack design VF CW (Gross et al., 2008).

Table 2.2 Summary of the different constructed wetlands reviewed in this chapter.

CW type	Add-on	Area m²	HLR L d⁻¹	HRT d	PE	Design Area per PE m² PE⁻¹	Influent COD mg L⁻¹	Effluent COD/RE mg L⁻¹ / %	Influent NH₄⁺-N mg L⁻¹	Effluent NH₄⁺-N/RE mg L⁻¹ / %	Influent TN mg L⁻¹	Effluent TN/RE mg L⁻¹ / %	Ref.
(1)	(2)	(3)	(4)	(5)	(6)	(7)	(8)	(9)	(10)	(11)	(12)	(13)	(14)
Aeration and Recirculation													
VF	None	33.3	2300	-	22.7	1.5	1365	669/51	239	193/19	-	-/-	(1)
VF	Recirculation	33.3	2100	-	14.9	2.2	1173	263/78	309	92/70	-	-/-	(1)
VF (tidal)	Passive aeration	0.1	31.2	-	0.0832ᶜ	1.2	-	-/-	40	2.3/94	-	-/-	(2)
HF	None	1	30	4	0.135ᵃ	7.4	540	-/> 89	-	-/-	16ᵈ	-/70ᵈ	(3)
HF	Continuous aeration	1	30	4	0.135ᵃ	7.4	540	-/94	-	-/-	16ᵈ	-/88ᵈ	(3)
SF	None	0.866	11.5	11	0.082ᶜ	10.6	-	-	107	54/51	-	-/-	(4)
SF	Continuous aeration	0.866	11.5	11	0.082ᶜ	10.6	-	-	106	7/93	-	-/-	(4)
VF	None	2.25	155	~0.25	-	3.6	438	88/80	58	12/79	73	52/29	(5)
VF	Intermittent Recirculation	2.25	378	~0.25	-	1.4	438	68/85	58	16/72	73	42/43	(5)
VF	Intermittent Aeration	2.25	356	~0.25	-	1.8	438	52/88	58	20/66	73	37/49	(5)
VF	Intermittent aeration and recirculation	2.25	403	~0.25	-	1.5	438	37/92	58	12/79	73	18/75	(5)
VF	None	0.018	13.4	0.5	0.016	1.1	158	48/48	10.6	5/40	12.7	8/20	(6)
VF	Continuous aeration	0.018	13.4	0.5	0.016	1.1	158	24/72	10.6	2.8/65	12.7	7.5/27	(6)
VF	Intermittent aeration	0.018	13.4	0.5	0.016	1.1	158	32/64	10.6	3.5/55	12.7	7.4/30	(6)

Table 2.2 (Continued)

CW type	Add-on	Area	HLR	HRT	PE	Design Area per PE	Influent COD	Effluent COD/RE	Influent NH$_4^+$-N	Effluent NH$_4^+$-N/RE	Influent TN	Effluent TN/RE	Ref.
		m²	L d⁻¹	d		m² PE⁻¹	mg L⁻¹	mg L⁻¹ / %	mg L⁻¹	mg L⁻¹ / %	mg L⁻¹	mg L⁻¹ / %	
(1)	(2)	(3)	(4)	(5)	(6)	(7)	(8)	(9)	(10)	(11)	(12)	(13)	(14)
Tidal constructed wetlands													
Tidal	Recirculation	0.03	13.3	0.21	0.30	0.10	2464	-/77	121	-/62	-	-/-	(7)
Tidal	None	0.04	64	0.21	3.4	0.01	4254	-/85	158	-/39	-	-/-	(8)
Tidal	Anti-sized	0.04	60.3	-	3.2	0.01	4728	-/74	159	-/33	-	-/-	(9)
Tidal	None	0.04	60.3	-	3.2	0.01	4728	-/71	159	-/34	-	-/-	(9)
Tidal	None	-	-	-	-	0.5	1298	-/84	176	-/93	222	-/78	(10)
Tidal	None	0.03	13.0	0.21	0.47	0.07	-	-/-	104	76/27	-	-/-	(11)
Tidal	Recirculation	0.03	13.0	0.21	0.29	0.11	-	-/-	121	63/48	-	-/-	(11)
Tidal	None	40	4804	-	160	0.25	2750	557/80	201	84/58	-	-/-	(12)
Constructed wetlands without pre-treatment													
VF	-	-	-	-	-	2.0	651	50/92	-	-/-	56ᵈ	7/90ᵈ	(13)
Stacked arrangement													
HF + VF	Passive aeration (cascade)	134.5	2152	5.4	57.4ᵃ	2.3	320	47/85	46	8.5/82	69	12.4/82	(14)
Towery HF + VF	Passive aeration (cascade)	134.5	4304	2.7	114.8ᵃ	1.2	320	49/85	46	7.7/83	69	11.9/83	(14)
Towery VF	Recirculation	0.21	30	1	0.14	1.5	358	91/33	-	-/-	46	25/45	(15)
VF	Recirculation	1	300	0.25	0.89	1.1	200	18/91	29	1.3/96	36	27/25	(16)
VF-VF	Passive aeration	263	3×10⁶	-	175	1.5	896	85/-	-	11/83	102ᵈ	13/84ᵈ	(17)
VF-HF	None	1.44	53.3	-	0.83ᵃ	1.7	2800	43/98	168	1.9/-	-	-/70	(18)
VF-HF	Recirculation	1.44	80.6	-	1.50ᵃ	1.0	3342	83/97	302	-/99	-	-/85	(18)
VF-HF	Recirculation	4	480	-	2.3ᵃ	1.8	1143	143/86	344	47/85	408	97/75	(19)
VF-HF	Recirculation	0.5	40	0.13(VF)-0.44(HF)	0.23ᵃ	2.2	1034	-/~48	448	-/~56	-	-/~58	(20)
HF-VF	Recirculation	0.5	40	0.13(VF)-0.44(HF)	0.23ᵃ	2.2	1034	-/~60	448	-/~57	-	-/~56	(20)

12

Table 2.2 (Continued)

CW type	Add-on	Area	HLR	HRT	PE	Design Area per PE	Influent COD	Effluent COD/RE	Influent NH_4^+-N	Effluent NH_4^+-N/RE	Influent TN	Effluent TN/RE	Ref.
		m^2	$L\ d^{-1}$	d		$m^2\ PE^{-1}$	$mg\ L^{-1}$	$mg\ L^{-1}/\%$	$mg\ L^{-1}$	$mg\ L^{-1}/\%$	$mg\ L^{-1}$	$mg\ L^{-1}/\%$	
(1)	(2)	(3)	(4)	(5)	(6)	(7)	(8)	(9)	(10)	(11)	(12)	(13)	(14)
Constructed wetlands on roofs													
HF	Intermittent aeration	6	480	2.1	0.16	37.5	87	19/78	-	-/-	-	-/-	(21)
HF	Intermittent aeration	6	480	2.1	1.31	4.6	495	159/68	-	-/-	-	-/-	(21)
Constructed wetlands on walls													
VF	None	1.2ᵉ	360	0.13	0.77	1.6	241	29/88	-	-/-	12.7	8.4/34	(22)
VF	None	0.04ᵉ	3	6	0.006	5.3	286	-/>81	8	0.1/99	40	-/>31	(23)
VF	None	0.04ᵉ	3	3.8	0.007	5.8	263	-/>83	22	-/>64	65	-/>50	(24)

CW, constructed wetland; HLR, hydraulic loading rate; HRT, hydraulic retention time; COD, chemical oxygen demand; TN, total nitrogen; PE, population equivalent; RE, removal efficiency; Ref., Reference; VF, vertical flow; HF, horizontal flow; SF, surface flow.

If the reference reported only a range of values (e.g. for HLR or pollutants concentrations) always the maximum values are reported in this table and also used for necessary calculations.

Values in Column (6) were calculated using 1 PE = 60 g BOD_5 d^{-1}, unless otherwise stated. When BOD_5 was not available, it was obtained from COD result using COD/BOD_5 = 2 and the letter ᵃ was displayed. Only for Kantawanichkul et al. (2001) and Kantawanichkul and Somprasert (2005), COD/BOD_5 = 3 as shown in Kantawanichkul et al. (2003) using the same wastewater. However, if neither BOD_5 nor COD were available, 1 PE = 15 g TKN d^{-1} was used. When TKN was not available, TN was displayed. If TN was also not available, NH_4^+-N was used and letter ᵇ was displayed. If all values were available, preference was always given to 1 PE = 60 g BOD_5 d^{-1} in this table for unification of the displayed results.

If values were displayed in Column (6), the values in Column (7) were also calculated as Column (3) ÷ Column (6). When no values were displayed in Column (6), values in Column (7) were taken directly from the reference.

ᵈ TKN.

ᵉ Horizontal area.

References: (1) Sun et al. 2003; (2) Green et al. 1997, 1998; (3) Ouellet-Plamondon et al. 2006; (4) Jamieson et al. 2003; (5) Foladori et al. 2006; (6) Dong et al. 2012; (7) Zhao et al. 2004a; (8) Zhao et al. 2004b; (9) Zhao et al. 2004c; (10) Zhao et al. 2011; (11) Sun et al. 2005; (12) Sun et al. 2006; (13) Sun et al. 2006; (14) Ye and Li 2009; (15) Zapater et al. 2011; (16) Sklarz et al. 2009, 2010; (17) Troesch et al. 2010; (18) Kantawanichkul et al. 2001; (19) Kantawanichkul et al. 2003; (20) Kantawanichkul and Somprasert 2005; (21) Winward et al. 2008; (22) Svete 2012; (23) Bazargan 2014; (24) Pérez Rubi 2014.

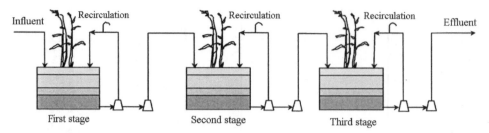

Figure 2.1 Schematic representation of the three-stage vertical flow constructed wetlands with recirculation. Taken from Sun et al. (2003).

2.2.2. Passive and active aeration

Vertical aeration pipes embedded in the CW bed have been used as another way to passively enhance oxygen availability for biodegradation and nitrification (Brix and Arias, 2005 - Fig. 2.2A; Troesch et al., 2010 - Fig. 2.6C). For instance, Green et al. (1998, 2006) developed a system using passive aeration by means of a pipe inlaid in a VF filter bed and a high-level siphon as the outlet (Fig. 2.2B). Wastewater flows through the bed, receives the treatment and finally is drained by means of the siphon (fill and drain operation system). A fast drainage assures that fresh air is sucked from the entrance of the pipe and the filling of the bed assures that the oxygen depleted air is evacuated through the pipes. This design to improve nitrification required approximately 1.2 m^2 PE^{-1} and reached an effluent concentration of 0-2.3 mg NH_4^+-N L^{-1} (Table 2.2) from an influent of up to 60 mg NH_4^+-N L^{-1}. However, the passive aeration method did not provide enough oxygen to treat higher NH_4^+-N influent concentrations.

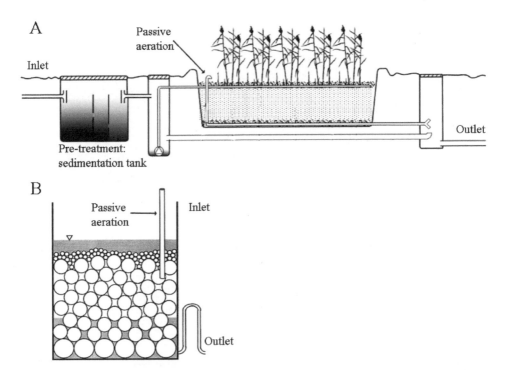

Figure 2.2 Schematic representations of passively aerated vertical flow treatment systems. Adapted from (A) Brix and Arias (2005) and (B) Green et al. (1998).

As passive aeration is usually not sufficient to meet the oxygen requirements for the treatment of many pollutants, the use of active aeration (also called artificial aeration or Forced Bed Aeration[TM] - following the Patent US 6,200,469 B1 from Wallace, 2001; Fig. 2.3A) started to gain popularity in the 90's in North America (Kadlec and Wallace, 2009). Artificially aerated systems typically require less footprint than conventional (non-aerated) CWs, thus, less material, workers and time is needed for its construction; in other words, less cost. However, the aeration equipment (e.g. air blower, pipes and emitters) and its maintenance will introduce some extra costs. Therefore, the active aeration investment is worth only when the capital cost is reduced and/or when the desired effluent quality, unachievable with a non-aerated system, is reached (Kadlec and Wallace, 2009). According to Kadlec and Wallace (2009), "a relatively gentle aeration" can deliver approximately 50-100 g O_2 m^{-2} d^{-1}, while non-aerated HF CWs can provide 1-6 g O_2 m^{-2} d^{-1}. Artificially aerated CWs usually show a more than 10-fold increase of removal rates compared to passive systems (Nivala et al., 2013).

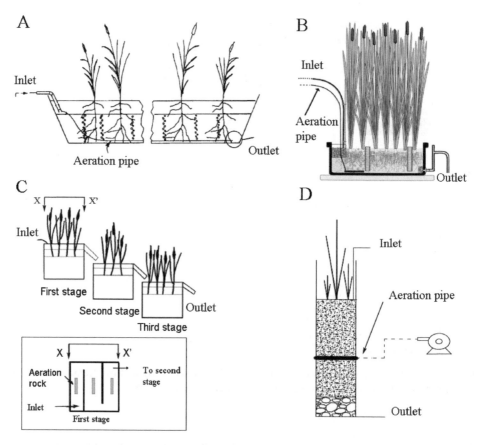

Figure 2.3 Schematic representations of different constructed wetlands with active aeration. Adapted from (A) Wallace (2001, Patent US 6,200,469 B1), (B) Ouellet-Plamondon et al. (2006), (C) Jamieson et al. (2003) and (D) Dong et al. (2012).

Artificial aeration was introduced to some HF CW mesocosms (designed for 7.4 m^2 PE^{-1}, Table 2.2) placed in a greenhouse treating trout farm effluent by means of a diaphragm air pump diffuser at the bottom of the bed at a rate of 2 ± 1 L min^{-1} (Ouellet-Plamondon et al., 2006; Fig. 2.3B). This study compared the benefits of aeration during summer and winter. The authors found a slightly higher pollutants removal in the aerated (94%) than in the non-aerated (89%) systems (Table 2.2), mainly during winter since plants are dormant and no oxygen translocation occurs. This little observed difference was due to the low organic loading rate (OLR) and the consequent large CW footprint (7.4 m^2 PE^{-1}) and probably, a higher OLR (and a consequent lower footprint) would have resulted in more visible effects of the artificial aeration.

Three aeration rocks were placed in the first bed of a three-stage free water surface CW, giving an aeration capacity of 5.5 L min^{-1} to enhance ammonia nitrogen removal (Jamieson et al., 2003; Fig. 2.3C). The system was designed for approximately 10.6 m^2 PE^{-1} (Table 2.2) and was tested for nitrification capacity with and without aeration. Results revealed that without aeration NH$_4^+$-N removal was only 51%, while aeration increased it till 93% (Table 2.2).

Dong et al. (2012) studied in China the possibility of using CWs to treat heavily polluted river water. Considering that China has limited land availability, it was intended to use a high hydraulic loading rate (HLR). The lower the HLR the better the pollutant removal but that implies a larger systems footprint. In those circumstances the use of aeration becomes necessary (Dong et al., 2012). They studied three VF CWs (Fig. 2.3D) each with a different aeration strategy (i.e. none, continuous and intermittent) and a design of 1.1 m^2 PE^{-1}. The continuous and intermittent aerated systems performed similarly, although the first achieved the highest nitrification rate (Table 2.2). However, considering operational costs as well, they concluded that intermittent aeration is the optimal mode for the VF CWs.

Foladori et al. (2013) tested VF CWs without aeration, intermittent aeration and combining intermittent aeration and recirculation designed for 3.6, 1.8 and 1.5 m^2 PE^{-1}, respectively (Table 2.2). The use of aeration increased the removal efficiency of the chemical oxygen demand (COD) and total nitrogen (TN) from 80% and 29%, respectively to 88% and 49%, respectively when aeration was used. Furthermore, when aeration was combined with recirculation the performance improved (removal efficiency of 92% for COD and 75% for TN). No major differences were observed in NH$_4^+$-N removal (Table 2.2). It should be emphasized that the conventional CW used the double amount of area than the other systems.

2.2.3. Fill and drain, tidal or reciprocating

Tidal flow CWs are systems that by filling and draining wastewater in the bed, enhances the entrance of fresh air into the system. It can work within the same bed (Green et al., 1997), between two beds (US Patent 5,863,433: Behrends, 1999), with several stages either between CWs (Zhao et al., 2004a,b) or combining technologies (i.e. CW and treatment lagoons, US Patent 6,863,816 B2: Austin et al., 2005). To allow the necessary oxygen transfer, the beds must be filled and drained several times per day ($\sim$ > 6 times d^{-1}; Kadlec and Wallace, 2009). The different cycles (filling and draining) allow for different conditions in the bed (aerobic and anoxic/anaerobic) that promotes a highly diverse microbial biomass, without domination of any particular type (Behrends et al., 2001). Behrends et al. (2001) clearly expressed that reciprocating CWs can be > 3 m depth in order to increase the fill and drain rates and thus, to reduce the land area requirements.

Behrends et al. (2007) combined the reciprocating CW with surface flow CW and HF CW and found that basically the reciprocating CW treated the majority of the wastewater, thus there was no need for the other types of CWs as they can stand alone. Zhao et al. (2011) showed that reciprocating CWs use 0.5 m^2 PE^{-1}. This CW concept, mentioned in Section 2.1.2 (Green et al., 1998), was first reported by Green et al. (1997) in Israel. Soon after many studies focused in this technology, mainly for high strength wastewaters like piggery effluent (Zhao et al., 2004a,b; Sun et al., 2005, 2006).

Zhao et al. (2004b) treated diluted pig slurry in a five-stage (laboratory scale) tidal flow CW (0.01 m^2 PE^{-1}; Table 2.2; Fig. 2.4A). Three operation conditions were tested and for all, a new tide was applied every 4 h, giving a HLR of about 1.6 m^3 m^{-2} d^{-1} for each stage CW. However, the saturated (fill)-unsaturated (drain) time (i.e. 3-1, 2-2 and 1-3 h per operation cycle) and the hydraulic retention time (HRT) (i.e. 15, 10 and 5 h d^{-1}) varied. The systems rested one week after each week of operation. The operational condition with the shorter saturation time (1 h) and shorter HRT (5 h d^{-1}) resulted in higher total removal efficiencies (~38-85%) than the other operational conditions (~25-79%), as the aeration by convection and diffusion during the unsaturated time in the tidal flow CW played a role in the pollutants removal of this high strength wastewater. A longer unsaturated time draws a higher oxygen flux into the CW beds. Phosphorus removal was independent of the operational conditions applied. Similar removal efficiencies (45-82%) were obtained by Sun et al. (2006) using a five-stage pilot scale tidal flow CW and treating 0.12 m^3 m^{-2} d^{-1} of a similar wastewater (diluted pig slurry) designed to use an area of 0.25 m^2 PE^{-1} (Table 2.2, Fig. 2.4B).

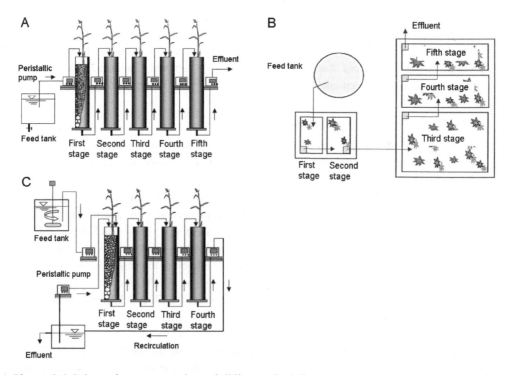

Figure 2.4 Schematic representation of different tidal flow constructed wetlands. Adapted from (A) Zhao et al. (2004b), (B) Sun et al. (2006) and (C) Zhao et al. (2004a) and Sun et al. (2005).

The use of recirculation in tidal flow CWs has also been successfully tested (Zhao et al. 2004a; Sun et al., 2005) with high strength wastewater (Fig. 2.4C). Recirculation, as seen in Section 2.1.1, prolongs the wastewater-biofilm contact that combined with the oxygen benefits of the tidal flow CW, provide an extensive removal of organic matter and NH_4^+-N. Zhao et al. (2004a) showed a COD and NH_4^+-N removal efficiency of 77% and 62% using an area of 0.1 m^2 PE^{-1} (Table 2.2). The study of Sun et al. (2005) in Table 2.2 showed that NH_4^+-N removal is improved by applying recirculation (from 27 to 48%). BOD_5 and total suspended solids (TSS) also improved (from 57 to 75% and from 57 to 68%, respectively), while the area demand was very similar (difference of only 0.04 m^2 PE^{-1}, Table 2.2).

Zhao et al. (2004c) developed a new CW design called the "anti-sized" CW. They arranged the substrate layers with coarse grains at the surface and fine grains at the bottom and worked under a tidal flow regime (with 5 stages) (similar to Figure 2.4C but with opposite grain arrangement). This new arrangement aids with the convection and diffusion of air into the bed for more rapid bio-decomposition and mineralization of removed TSS and increases the soil storage capacity allowing the bed matrices to be more efficiently used. It was shown that the "anti-sized" CW delays the clogging process occurrence while contaminants (COD, BOD_5, NH_4^+-N, phosphorus and TSS) were similarly removed as in the "progressively-sized" (finer grain at the top and coarser grain at the bottom) CW. Both systems were designed for 0.01 m^2 PE^{-1} (Table 2.2).

2.2.4. **No pre-treatment**

The general design of a CW always includes a pre-treatment (e.g. septic tank, digesting tank and settling basin, Fig. 2.2A) (Brix and Arias, 2005; Kadlec and Wallace, 2009); avoiding the pre-treatment would enhance the organic load entering the CW, thus decreasing the area needed per PE. The classical "French systems" consist of a two-stage VF CW fed directly with raw wastewater (Molle et al., 2005; Troesch et al., 2014). The first stage consists in three identical beds that have a total area of 1.2 m^2 PE^{-1} and the second stage consists in two identical filters with a total area of 0.8 m^2 PE^{-1} (total filtration area of 2 m^2 PE^{-1}, Table 2.2, Fig. 2.5). The key in the success of such systems is the alternate feeding-resting regime applied: each primary stage bed has a HRT of 3.5 d and a resting period of approximately twice the feeding time (Troesch et al., 2014). Consequently, the attached biomass on the filter media can be controlled and the deposited sludge (on the first stage) is able to mineralize (Troesch et al., 2014). The system called Bi-filtre® was as well designed to avoid the use of a pre-treatment (Section 2.2).

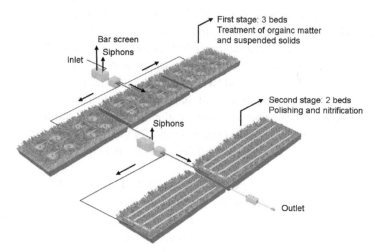

First stage: 3 beds
Treatment of orgainc matter
and suspended solids

Bar screen
Siphons
Inlet

Second stage: 2 beds
Polishing and nitrification

Siphons

Outlet

Figure 2.5 Schematic representation of the classical French System (two-stage vertical flow constructed wetlands) treating raw domestic wastewater. Taken from Troesch et al. (2014).

2.2.5. Enhancing phosphorus removal

CWs are not effective in removing phosphorus (Prochaska et al., 2007). To increase the CW efficiency aiming for phosphorus removal, commonly a larger system is required that can provide more adsorption sites (Kadlec and Wallace, 2009). This practice results in a permanent larger footprint that offers only a temporarily extra phosphorus removal. It is then necessary to boost the phosphorus removal by CWs without enlarging the system footprint.

In CWs, the main phosphorus removal mechanisms are precipitation and adsorption. Thus, a cost-effective and, to some extent, efficient solution for CWs is the use of a filter medium with high adsorption capacity and high content of the cations that are able to precipitate phosphorus (e.g. Fe^{3+}, Fe^{2+}, Al^{3+} and Ca^{2+}).

Different materials (natural, industrial by-products and man-made products) have been tested as substrate in CWs that increase the adsorption and precipitation rate, such as apatite, dolomite, opoka, maerl, shells, fly ash, iron ore, blast furnace, Filtralite P™ and crushed marble (Vohla et al., 2011). Depending on the phosphorus-removal media available, different alternatives can be implemented in the CW. If the reactive media is suitable to maintain proper hydraulic conditions of the CW (even when biofilm is attached) then it can be used as the filter medium. If not, to avoid the risk of clogging and to enhance an easy manipulation, a pre- or post-treatment should be used (Arias et al., 2001).

Another method to remove phosphorus from wastewater is the iron-contactor technique. The iron-contactor is a simple and low-cost approach for phosphorus removal from wastewater that does require neither chemicals nor additional equipment (US Patent 4,029,575: Bykowski and Ewing, 1977; Haruta et al., 1991). An iron-contactor is defined as a piece of iron, or alloys, submerged in a wastewater treatment tank and allowed to corrode. Due to corrosion, iron ions will elute from the iron-contactor surface and will combine with the phosphorus compounds dispersed in the wastewater in order to form insoluble iron-phosphorus compounds that will be removed physically. It was estimated that the corrosion rate is about 1 mm or less in 30 yr, assuring long duration of a relatively thick iron piece (Haruta et al., 1991). It has been shown that the iron-contactor technique does not interfere negatively during removal of BOD_5 and TSS, neither in the processes of nitrification-denitrification (Haruta et al., 1991; Choung and Jeon, 2000). The idea of the iron-contactor has been used already in CWs, using iron fillings in the media (Lüderitz and Gerlach, 2002) instead of a separate tank. The addition of iron fillings into the CW bed is more effective in removing phosphorus than the use of Ca-rich soils (Lüderitz and Gerlach, 2002). They studied three systems (VF, HF and circular HF) with a specific filter volume design of 4, 7 and 7 m^3 PE^{-1}, respectively. Only the HF included iron fillings. They showed that TP removal efficiencies ranged from 27-44%, 97-99% and 38-72%, respectively.

Nanotechnology also offers options for phosphorus removal. Oxides of polyvalent metals exhibit favorable ligand sorption properties and from them, hydrated ferric oxides (HFO, size 10-100 nm) are innocuous, inexpensive, readily available, and chemically stable over a wide pH range. If those are dispersed within polymeric anion exchangers, the sorption capacity for phosphorus increases. Blaney et al. (2007) developed the HAIX (hybrid anion exchanger) for the selective removal of phosphorus by doping HFOs within a strong-base anion exchange resin. The HAIX is able to capture phosphates for quite a long period of time and after saturation, a simple NaOH-NaCl washing regenerates and recovers over 95% of its capacity. A similar technology, HFO-201, was later developed using a polymeric anion exchanger called D-201 (Pan et al., 2009). Initial phosphorus concentrations in synthetic and real wastewater for the HAIX study was 260 µg L^{-1}, while for the HFO-201 study it was 2 mg L^{-1}. Final concentrations were less than 50 and 10 µg L^{-1}, respectively. To the best of the author's knowledge, no combination of nanoparticles with CWs has been studied yet.

2.3. STACKING UP EXTRA TREATMENT STAGES

Increasing the total system footprint by constructing several treatment stages next to each other can be substituted by placing each stage on top of the other (stack up design), thus increasing the system depth, but not the area demand.

Ye and Li (2009) designed the towery hybrid CW (three circular cells one on top of the other) (Fig. 2.6A). The influent domestic wastewater was added to two different zones (80% to the inlet zone and 20% to the tower zone) and all would be combined at the outlet zone. The towery hybrid CW was developed to increase the nitrification rate by elevating the dissolved oxygen (DO) level and maintaining oxic conditions in the wastewater by passive oxygenation due to a cascade-type current starting in the uppermost cells (Ø 3 m), flowing to the middle cell (Ø 5 m) and falling to the bottom cell (Ø 7 m). Although its purpose was not the footprint reduction, the towery hybrid CW used the principle of stacking up different stages to save land area. They proved that their new design enhanced organic matter degradation (inlet zone), nitrification (tower zone) and denitrification (outlet zone) processes (Fig. 2.6A). An area of approximately 1.2-2.3 m^2 PE^{-1} was used for the design after successfully (effluent COD and TN concentrations of < 50 and < 13 mg L^{-1}, respectively) testing two HLRs (16 and 32 cm d^{-1}).

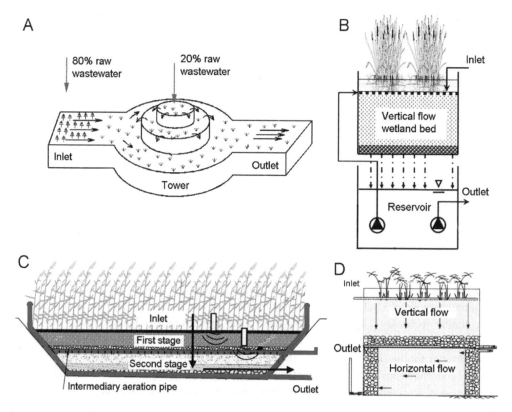

Figure 2.6 Schematic representations of different constructed wetlands using a stack up design. Adapted from (A) Ye and Li (2009), (B) Sklarz et al. (2010), (C) Troesch et al. (2010) and (D) Kantawanichkul and Somprasert (2005).

A group of researchers (Gross et al., 2007, 2008; Sklarz et al., 2009, 2010; Zapater et al., 2011) developed the recirculating VF CW consisting of two stacked containers: an upper container consisting of a VF CW and a bottom one working as a reservoir (Fig. 2.6B). Domestic wastewater enters the top container, passes through the bed and trickles down to the reservoir through perforated holes at the bottom of the top container. This practice enriches the water passive aeration as each drop falling into the reservoir is exposed to the atmosphere. A recirculation pump placed in the reservoir recycles back the water to the bed, repeating the water cycle until the desired water quality is achieved. The recirculating VF CW needs only 1.0-1.5 m^2 PE^{-1} (Table 2.2).

The Bi-filtre® was developed in France with the intention of reducing the system's footprint (Troesch et al., 2010). This deep system, patented (EP-042931295) by Epur Nature (France, www.epurnature.fr), consists of two vertical stages placed one on top of the other (2.7 m² PE⁻¹ in total, 1.5 m² PE⁻¹ and 1.2 m² PE⁻¹ on the upper and lower stage, respectively, but a filter footprint of 1.5 m² PE⁻¹, Fig. 2.6C; Table 2.2), holding *Phragmites australis* on the upper stage, treating raw wastewater (avoiding pre-treatment, see Section 2.1.4) and reaching removal efficiencies of > 85% for COD (effluent 85-125 mg L⁻¹), 83% for NH_4^+-N (effluent 11 mg L⁻¹) and 84% for TKN (effluent 13 mg L⁻¹) (Table 2.2).

A research group in Thailand investigated the combination of a VF CW on top of a HF sand filter for the treatment of pig farm wastewater (Kantawanichkul et al., 2001; Kantawanichkul et al., 2003; Kantawanichkul and Somprasert, 2005) (Fig. 2.6D). The study of Kantawanichkul et al. (2001) showed that the systems, with or without recirculation, can remove similarly the COD concentrations (97%, from influents of above 2800 mg L⁻¹, Table 2.2). Thus, recirculation did not enhance the treatment performance for COD, however for TN it was enhanced from 70% up to 85%. The design area was ~1.7 and 1 m² PE⁻¹ for the system without and with recirculation, respectively (Table 2.2). Kantawanichkul et al. (2003) and Kantawanichkul and Somprasert (2005) studied the same type of system but testing different HLRs. The first showed that increasing the HLR did not deteriorate the effluent quality (except when using 240 L d⁻¹, since the systems experienced clogging), while the second showed the opposite (Table 2.2). Thus, the design area of the systems used in Kantawanichkul et al. (2003) can be 1.8 m² PE⁻¹ while the area of Kantawanichkul and Somprasert (2005) should be higher than 2.2 m² PE⁻¹.

2.4. PLACING THE CONSTRUCTED WETLAND AT UNUSED SPACES

Employing commonly unused spaces, namely a roof or a wall, to locate the CW is another strategy to minimize the systems footprint. In such a scenario, the wastewater treatment will occur elsewhere and not on the land, hence implying the use of 0 m² PE⁻¹.

2.4.1. Roofs

To the best of the author's knowledge, few studies have addressed the challenge of wastewater treatment on a roof: a natural wastewater treatment system at the John Deere plant in Mannheim, Germany (Transfer, The Steinbeis Magazine, 2010), a Roof Garden study conducted by Anhalt University of Applied Sciences, Bernburg, Germany (Thon et al., 2010) and a novel subsurface CW system called Green Roof Water Recycling System (GROW) patented in 2004 by Water Works, London, UK (Avery et al., 2006; Frazer-Williams et al., 2006; Winward et al., 2008).

Little information concerning the design and performance of the systems has been published about the German studies. While the first study only mentioned that no-substrate is used in the system (only plants and wastewater); the second describes the study setup as "shallow horizontal filters", "miniature roof gardens" and "CWs on the roof" and confirms that CWs on the roof have the potential to positively affect the microclimate in dense urban areas (Thon et al., 2010). The third, GROW, is a NWTS for the treatment of greywater suitable for the use on sloped roofs. It comprises five rows of two troughs connected in series filled with a bottom 10 cm layer of Optiroc, a light expanded clay aggregate (LECA), to fulfill the weight requirements (50 kg m^{-2}) and a top 6 cm gravel chippings (10-20 mm diameter) with baffles and weirs within the media to force the flow through the whole of the media. The substrate is covered with reinforced membranes to prevent entry of rainwater, and aquatic marginal plants are inserted into the media through small slits in the membrane. Aeration is provided for 1 h d^{-1} and 480 L of water are pumped daily as continuous flow (0.07 m d^{-1}), having a HRT of 2.1 d. Such a system has proven that it is a competent system when treating low strength wastewater with removal efficiencies of approximately BOD 90%, COD 78%, TSS 90% and turbidity 96% (similar to those in common CWs) (designed for 37.5 m^2 PE^{-1}). However, it has a decreased performance when treating high strength wastewater with removal efficiencies of approximately BOD 51%, COD 68%, TSS 79% and turbidity 53% (designed for 4.6 m^2 PE^{-1}) (Winward et al., 2008; Table 2.2). Moreover, GROW showed a better performance than a conventional HF CW when removing indicator bacteria (total coliforms, *E. coli*, enterococci, clostridia) and *P. aeruginosa* from low strength wastewater.

Another attempt has been the use of Wetland Roofs since 2006 by Blumberg Engineers, a German company expert in ecological engineering (www.blumberg-engineers.de). They called Wetland Roofs a versatile and innovative type of green roof (GR) that contains helophytes (that were pre-cultivated in advance in basins) inlaid on mats of non-woven material. To assure the humidity necessary for the plants, artificial irrigation is installed on the roof pumping rainwater usually stored by cisterns. The systems are mainly used for the retention and purification of stormwater but, although it has not been proven yet, they claim it can work for the treatment of domestic, agricultural and industrial wastewater.

2.4.2. Walls

Green walls (also called living walls or life panels) for wastewater treatment have been studied as well. However, the majority of studies were done only as MSc research in The Netherlands (Ahmed, 2012; Bazargan, 2014; Pérez Rubí, 2014), Belgium (Bäumer, 2013) and Norway (Svete, 2012); some using greywater and others primary treated domestic wastewater.

Svete (2012) used green wall modules divided in 3 sections and all together treated 360 L d^{-1} (Fig. 2.7A). The wall had approximately 2 m height and a removal capacity of > 90% for TSS, > 82% for COD, > 95% for BOD$_5$, > 69% for total phosphorus (TP) and > 31% for TN. The effluent concentrations were lower than 4, 43, 6, 0.36 and 8.8 mg L^{-1}, respectively. The total area used for all 3 sections was 1.2 m^2 (Table 2.2). It was claimed that a filter surface area of ~1 m^2 would serve to treat the grey water from a 4 person household.

Bazargan (2014) and Pérez Rubí (2014) showed satisfactory treatment of pre-settled domestic wastewater using green walls with similar dimensions (0.2 m width × 0.2 m length × 0.5 m depth) but packed with different substratum that generated a different HRT (6 and 3.8 d, respectively) for the same HLR (~3 L d^{-1}). Organic matter (> 80% for BOD$_5$ and COD, effluent < 65 mg COD L^{-1} and < 47 mg BOD$_5$ L^{-1}) and solids (> 87% removal of TSS, effluent < 6 mg L^{-1}) removal were high (Fig. 2.7B) (Table 2.2). For nutrients, as commonly found in CWs, the removal was much lower (> 31% for TN and > 39% for TP; effluent < 51 mg TN L^{-1} and < 12 mg TP L^{-1}). Both studies showed that the surface area necessary in the green wall design would be approximately 10 m^2 PE^{-1}. However, this area is attached to a building wall and thus, it is considered that almost no space was taken for the wastewater treatment. The (horizontal) wetted surface area taken by a green wall module was of 0.04 m^2, and using this value, the designed space would be of ~5.3-5.8 m^2 PE^{-1} (Table 2.2).

The company EcoWalls LLC (New Jersey, USA; www.greenecowalls.com) specialized in promoting green infrastructures mentioned that EcoWall® can act as a compact filtration system with a smaller footprint than conventional CWs. However, no further details about a green wall project for wastewater treatment is provided in their website. Also, the company GSky Plant Systems Inc. (Florida, USA; www.gsky.com) is a leading provider of vertical Green Walls in North America and the Middle East. Among their projects, one was found at the Bertschi School built in 2011 that mentions that it can "filter" all grey water in the building (Fig. 2.7C). However, no further water quality information is available in their website.

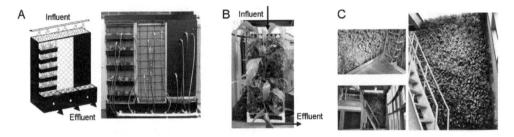

Figure 2.7 Green walls for wastewater treatment. Taken from (A) Svete (2012); (B) Pérez Rubí (2014) and (C) Gsky Plant Systems Inc. website - Bertschi School Project.

2.5. EVOLUTION OF THE CONSTRUCTED WETLAND FOOTPRINT OVER TIME

Table 2.3 shows the area requirement of different types of CWs. It contained the information displayed in Table 2.2 and supplemented with extra references (organized by the publication year), in order to look for trends in the CW area requirement development in time. The earliest documented CW relates to a patent from 1901, which called the system "purifying water" (US Patent 681,884: Monjeau, 1901). More research about (HF) CWs was developed in the 1950's in Germany, however it was not until the 1980's that this technology started to spread to other countries (Vymazal, 2010). This explains why the references included in Table 2.3 start only in the 1980's.

No trend towards increasing or decreasing the area demand of CWs is visualized from past to present (Table 2.3). In fact, the conventional CWs (HF, VF and hybrid) footprint has maintained since their early developments, despite scientific efforts for improving their performance. The only clear example for area reduction is the recently development of tidal CWs, that require areas of 0.01-0.5 m^2 PE^{-1} (Table 2.3), similar to the footprint required by other technologies shown in Table 2.1. However, in the case of CWs, the question of lifespan of the system due to clogging remains and must be considered during the design period.

Table 2.3 Area required by different constructed wetlands (CW).

CW type	Design area (m^2) per PE	Reference	Publication Year
HF	4.2	Radoux and Kemp 1982 (cited in Gómez Cerezo et al. 2001)	
HF	2-5	Kickuth 1984 (cited in Brix, 1987a)	1980's
HF	1.9-23	Brix 1987a	
HF	5	Brix 1987b	
VF filter (tidal)	1.2	Green et al. 1997, 1998	1990's
Hybrid	4.1-5.1	Brix and Johansen 1999	
Hybrid	0.8-2.3	Gómez Cerezo et al. 2001	
VF	1.5-2.2	Sun et al. 2003	
SF	10.6	Jamieson et al. 2003	
Tidal	0.01-0.1	Zhao et al. 2004a,b,c	
VF	0.3-3.1	Cooper 2005	
Tidal	0.07-0.11	Sun et al. 2005	2000's
Tidal	0.25	Sun et al. 2006	
HF	7.4	Ouellet-Plamondon et al. 2006	
HF	3-6	Albuquerque et al. 2008	
HF / VF / SF	5 / 1.5-6 / 20	EPA 2009	
Towery (hybrid)	1.2-2.3	Ye and Li 2009	
VF	1.1	Sklarz et al. 2009, 2010	
HF / VF	1 / 2	O'Hogain et al. 2011	
VF	1.5	Zapater et al. 2011	
Tidal	0.5	Zhao et al. 2011	2010's
VF	1.1	Dong et al. 2012	
VF	1.4-3.6	Foladori et al. 2013	
VF (French system)	2.0	Troesch et al. 2014	

PE, population equivalent; VF, vertical flow; HF, horizontal flow; SF, surface flow.

2.6. CONCLUSION

This chapter showed that many approaches have been undertaken in order to improve the efficiency of CWs and, indirectly, to reduce their land requirement with different success. However, the importance of saving area has not been specifically undertaken by the majority of the above mentioned literature. Thus, the combination of some of the other ideas (e.g. recirculation, aeration, specific filter media for phosphorus removal and stack design) should be further tested with an emphasis on the approach of saving area. Furthermore, the most evident way to reduce the space requirement (up to 0 m^2 of land PE^{-1}) is by placing the CW on unused spaces such as roofs and walls. Hence, this thesis will work through three different (mentioned) approaches by studying: (i) different materials for an enhanced phosphorus removal, (ii) different types of Duplex-CW (hybrid system in stack arrangement), testing different approaches such as recirculation and aeration, and (iii) a constructed wetroof, a CW placed on the roof of a building.

CHAPTER 3.

USE OF MARINE AND ENGINEERED MATERIALS FOR THE REMOVAL OF PHOSPHORUS FROM SECONDARY EFFLUENT

Constructed wetlands (CWs) require large area per population equivalent for the treatment of domestic wastewater. Hence, more compact CWs with equivalent or higher treatment efficiency per m^2 than conventional CWs need to be developed. The aim of this study was to reduce the required area by enhancing the phosphorus removal through the use of marine (i.e. crushed coral, oyster-shells and mussel-shells (raw and pyrolyzed)) and engineered (i.e. nanoparticle-beads) materials. This was done in batch and column experiments.

The pyrolyzed materials and the nanoparticle-beads showed a phosphorus removal capacity exceeding 99%, respectively through precipitation and adsorption. The conditions each material needed for the removal were different (e.g. contact times and material-to-solution ratios). Conversely, the raw marine materials did not achieve high removal efficiencies (12-59% after 7 d), unless the pH was increased to approximately 12. In general, all materials achieved phosphorus-removal levels beyond typical CW, the pyrolyzed materials and nanoparticle-beads being the most effective of the materials investigated. However, the high pH (~12) of the effluent after the treatment with pyrolyzed material can be a limitation of its application. A (separate) post-CW filter, packed with either pyrolyzed materials or nanoparticle-beads is proposed to increase the phosphorus removal efficiency thereby reducing the total space requirement of a CW. Recommendations for practical use are also included in this study.

This chapter is based on:
Zapater-Pereyra M., Malloci E., Bruggen van J.J.A., Lens P.N.L. (2014), "Use of marine and engineered materials for the removal of phosphorus from secondary effluent", Ecological Engineering, 73, 635-642.

3.1. INTRODUCTION

Phosphorus, mainly from untreated wastewater, has been identified as the main contaminant causing algal blooms and subsequent eutrophication in water bodies (Rittmann et al., 2011; Metcalf and Eddy, 2003). Concentrations as low as 100 μg L^{-1} still provide sufficient nutrients to cause eutrophication (Bitton et al., 1974). At the same time, phosphorus is a non renewable resource, well used in industry and agriculture, predicted to dwindle in the next fifty years (Sengupta and Pandit, 2011). Hence, phosphorus present in wastewater should be recovered and reused.

Constructed wetlands (CWs) are not usually effective in removing phosphorus (Prochaska et al., 2007; Arias and Brix, 2005; Lüderitz and Gerlach, 2002), seldom achieving residual phosphorus concentrations below the limits that can avoid eutrophication. Although CWs are recognized as an efficient natural wastewater treatment system, their poor phosphorus removal efficiency is still a hurdle for wider application.

The main removal mechanisms of phosphorus in CWs are plant uptake, precipitation and adsorption. The amount of phosphorus removed when harvesting the plants is small (2-4.9 g P m^{-2} yr^{-1}) as compared to the amount of phosphorus entering wetlands via wastewater (typically 150-300 g P m^{-2} yr^{-1}) (Kadlec and Wallace, 2009; Arias et al., 2001). Phosphorus can be precipitated by Mg^{2+}, Fe^{3+}, Fe^{2+}, Al^{3+} and Ca^{2+} cations to form insoluble compounds (Rittmann et al., 2011). In CWs, a filter medium with a high content of these cations can thus be used to enhance precipitation and to increase adsorption sites. However, results are not yet satisfactory: phosphorus is not removed sufficiently to avoid eutrophication; phosphorus recovery is rather difficult; and eventually media saturation will occur. To increase the CW efficiency aiming for phosphorus removal, commonly a larger system is used (more media and therefore, more adsorption sites). This may temporarily enlarge the system's life span but will introduce the disadvantage of a larger footprint. For these reasons, it is recommended to have a compact and external phosphorus post-treatment step that serves as a polishing function for a CW and also provides the opportunity to recover phosphorus without expanding the system size.

This study tests the suitability of marine and engineered materials for the post-treatment of phosphorus from secondary effluent (e.g. from CWs). The marine materials (e.g. oyster- and mussel-shells) are a renewable source of $CaCO_3$ and some are available on the seashore or as a waste product of shellfish farms (Abeynaike et al., 2011). The engineered material is a phosphate selective resin dispersed with iron oxide nanoparticles (iron content of 75-90 mg Fe g^{-1} resin), providing active adsorption sites for the removal of phosphates (Sengupta and Pandit, 2011).

Several tests were performed to understand the phosphorus removal mechanism of the materials investigated and to determine their capacity, mode of use and application in CWs. This study specifically aimed to: (i) assess the phosphorus removal efficiency of the selected materials, (ii) ascertain the phosphorus removal mechanism of the marine material (after pyrolysis) and (iii) propose an efficient phosphorus removal material (with some practical recommendations) that can improve the phosphorus treatment provided by CWs, thereby avoiding the costly need to oversize for sufficient phosphorus removal.

3.2. MATERIALS AND METHODS

3.2.1. Tested material and phosphorus source

The tested phosphorus-sorbing materials were oyster-shells (OS), mussel-shells (MS), crushed coral (CC) and nanoparticle-beads (NB). To the authors' knowledge, this is the first study using NBs as a potential phosphorus polishing material for CWs. All materials, except for the NB, were used in two forms: natural (called "raw") and pyrolyzed under nitrogen gas at 750 °C for 1-2 h (called "pyrolyzed") according to Kwon et al. (2004). The source of phosphorus used in the study was a mono-phosphate solution (~30 mg P L^{-1}, pH ~5, demineralized water spiked with KH_2PO_4) and domestic wastewater (~10 mg P L^{-1}, pH ~7, primary settled effluent from Harnaschpolder wastewater treatment plant, Delft, The Netherlands) (Table 3.1). It should be noted that the concentration used in the phosphate solution is above of what it is reported as high strength domestic wastewater in the literature (12 mg L^{-1} of total P, Metcalf and Eddy, 2003), but the intention was to assess the removal limits of the materials investigated.

Table 3.1 Physico-chemical composition of the raw domestic wastewater used in this study.

Parameter	Unit	Value
pH	-	7.3 ± 0.3
Temperature	°C	14.3 ± 2.2
DO	mg L^{-1}	2.9 ± 1.8
EC	µs cm^{-1}	1273 ± 411.8
BOD_5	mg L^{-1}	134.5 ± 69.9
COD	mg L^{-1}	435 ± 12.3
DOC	mg L^{-1}	74.7 ± 31.7
TSS	mg L^{-1}	135.4 ± 69.1
NO_3^--N	mg L^{-1}	0.3 ± 0.4
NH_4^+-N	mg L^{-1}	43.2 ± 17.5
PO_4^{3-}-P	mg L^{-1}	7.8 ± 2.8
E. coli	CFU $100mL^{-1}$	$6.2 \times 10^6 \pm 2.2 \times 10^6$
Total coliform	CFU $100mL^{-1}$	$22 \times 10^6 \pm 8.4 \times 10^6$
Alkalinity	mg L^{-1} as HCO_3^-	428.2 ± 13.8
Ca^{2+}	mg L^{-1}	65

3.2.2. Batch experiments

Batch experiments were conducted in duplicate using 500 mL plastic bottles containing a phosphorus source and 10 g of material. The bottles were capped and placed immediately in a shaker at 100 rpm for 7 d. For comparison reasons all the materials, except for the NB, were crushed to the same size (< 0.3 mm). In some batch tests, silica sand (1-1.6 mm) was added to simulate a mixture with the CW matrix since sand is a substrate widely used in CWs. Depending on the batch test (with or without sand), one or two control bottles were included:

(i) control, 500 mL of phosphorus source and (ii) control-sand, 450 mL of phosphorus source plus 150 g of sand. Each experiment consisted of the following sequential steps: filling the 500 mL plastic bottle with the phosphorus source, addition of sand (when required by the experiment), addition of the tested material, placement in the shaker and reaction period. The phosphate concentration and pH were measured and sampling was conducted at: T_0 (time zero, solution/wastewater alone), T_S (time zero-sand, immediately after the sand was added, if added), T_M (time zero-material, immediately after the material was added) and after 1 h, 1 d and 7 d.

The influence of the high pH on the raw marine material was tested. For that, a similar duplicate batch experiment was conducted using the phosphate solution mixed either with pyrolyzed or high-pH (~12, adding 10 M NaOH) raw marine material. Two control bottles were used: (i) control (phosphorus source only) and (ii) control-pH (phosphorus source with high pH). The sampling regime was conducted at T_0, $T_{M \; \& \; adjusted \; pH}$ (time zero-material immediately after the pH modification) and after 1 h, 1 d and 15 d.

3.2.3. Column experiments

Sand, NB and OSs (pyrolyzed and raw) were selected to build nine columns of 2.5 cm internal diameter and 15-30 cm depth according to Table 3.2. To avoid cementation in the OS columns, the packing material consisted of 3-5 mm beads and was manually prepared: pyrolyzed material and water were mixed (1:1 ratio) to form a moldable paste. After pyrolyzation, the material was sieved and only the size range of 1-3.2 mm was used to form the paste. Such size range was optimal to form the required paste. The paste was placed in a piping-bag and then pressed to form the beads on a flat surface (Fig. 3.1). The beads were let to dry (ambient temperature) for at least 24 h.

Figure 3.1 Preparation of pyrolyzed oyster-shell beads with the paste and the piping-bag.

The columns were saturated with domestic wastewater in a downflow mode and the outlets were adjusted to a flow rate of 0.5 mL min^{-1}. Between the material and the outlet pipe, a piece of glass wool was used to avoid loss of the material. Influent and effluent water samples were analyzed for pH and PO_4^{3-}-P during 2 months (daily and weekly during the first and second month, respectively): 7 d of startup period where diluted (5:1) domestic wastewater was applied and 54 d of experimentation period where raw domestic wastewater was used. Ca^{2+} ions were measured randomly in the raw (n = 4) and pyrolyzed (n = 8) material effluent.

Table 3.2 Characteristics of the columns used in this study.

	Pyrolyzed oyster-shell beads	Raw oyster-shell	Nanoparticle beads	Nanoparticle beads + sand	Sand
Number of columns	2	2	2	2	1
Bed height (cm)	15	15	15	30 (15 + 15)	15
Mass of media (g)	65	100	70	220 (70 + 150)	150
Particle size range (mm)	3-5	1-3.2	0.5-2	0.5-2 + 1-1.6	1-1.6

3.2.4. Tests for pyrolyzed material phosphorus removal mechanism

The phosphorus removal mechanism of OS was tested by the Imhoff cone test (APHA, 2005) and the leaching test (norm 12457-2 from UNI EN, 2004). Two other tests were performed to further study the phosphorus removal mechanism. The first, the "solids/colloids experiment", consisted of adding 16.7 g of pyrolyzed OS (< 0.3 mm) to two beakers containing 500 mL of demineralized water (material-to-solution ratio of 1:30), which were stirred for 1 h. One mixture was filtered (0.2 μm), the solids discarded and the liquid phase was kept. To both beakers, 1 L of phosphate solution was added and stirred for 1 h. Sampling was conducted after filtration, immediately, 30 min and 1 h after the addition of the phosphate solution. This experiment was done to evaluate whether the presence (or not) of solids and colloids (pyrolyzed OS) contributed to the removal of phosphorus.

To test if the phosphorus removal efficiency is linked to a high pH (needed for precipitation), a second test, the "titration experiment", was performed. For this experiment, the pyrolyzed OS was mixed with water to form a water-OS solution (0.25 g OS : 1 mL water). Sequential additions of 1 mL water-OS solution and drops of 0.5 M H_2SO_4 (to maintain the pH below 8, to avoid precipitation) were added to a beaker containing 1 L of phosphate solution. The water-OS solution was used to provide an instant mixing during the titration. The pH was constantly recorded while the solution was stirred. A similar trial was conducted without acid.

Additionally, pure CaO was also titrated (0.2 g each addition, without acid) into a phosphate solution to compare the behavior with that of the pyrolyzed OS (without acid). Also, a phosphorus equilibrium solubility diagram was plotted to understand the effect of the pH on the concentration of the phosphorus removed by pyrolyzed OS. The construction of the diagram was done by mixing 1 L of phosphate solution, 1 g of pyrolyzed OS (< 0.3 mm) and drops of H_2SO_4 to decrease the pH from ~12 (achieved by the pyrolyzed OS) to approximately 2. The pH was constantly recorded and once the value was stable, a sample was taken for phosphorus analysis.

3.2.5. **Analytical methods**

The water samples were analyzed for pH by the electrometric method and PO_4^{3-}-P by the ascorbic acid method according to Standard Methods (APHA, 2005). Ca^{2+} ions were measured using an AAnalystTM 200 Atomic Absorption System-Flame (PerkinElmer, USA).

3.3. RESULTS

3.3.1. **Batch experiments**

3.3.1.1. **Marine materials**

The pyrolyzed marine materials, in the presence or absence of sand, showed similar trends when exposed to both phosphorus sources (Fig. 3.2). All pyrolyzed materials removed immediately (T_M) almost all the phosphorus (> 95%), and throughout the experiment the removal efficiency increased to (or maintained at) above 99%. In contrast, the raw material phosphorus removal at T_M was in the range of 12-17% for the wastewater (Fig. 3.2C) and 23-27% for the phosphate solution (Fig. 3.2A). When sand was added, the removal was higher when using wastewater (21-29%, Fig. 3.2D), but it slightly decreased for the phosphate solution (17-20%, Fig. 3.2B). In the control-sand, no immediate phosphate removal occurred (Fig. 3.2B and D, between T_0 and T_S). Throughout the experiments, the raw material never reached values similar to the ones achieved by the pyrolyzed material. Nonetheless, within 7 d, removal values increased to approximately 59% (raw OS), 49% (raw CC) and 46% (raw MS) for the phosphate solution (with and without sand). With wastewater, a marked difference was visible in the presence and absence of sand (Fig. 3.2): when sand was present, the phosphorus removal obtained by all the raw marine material (55%) was much higher than the removal without sand (17%). It should be noted that also the control-sand (with wastewater) decreased the phosphorus levels (Fig. 3.2D).

The mixing of the raw marine material with the phosphate solution (with and without sand) resulted in a variation of the pH to levels between 7.4-10.1; whilst in the case of wastewater, the pH was maintained in the neutral ranges (Table 3.3). Moreover, the absence of sand showed a slight instability in the pH, especially for that given by the OSs (8.7-10.1 and 8.0-8.3, without and with sand, respectively). With the few cases of pyrolyzed CC without sand and pyrolyzed MS with sand, the pyrolyzed marine material raised the pH to highly alkaline levels of approximately 12 (Table 3.3).

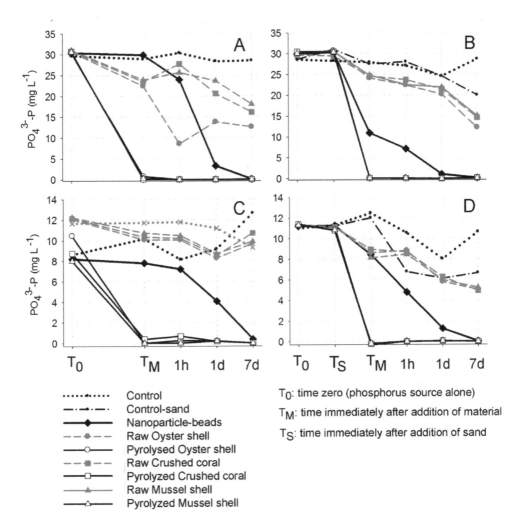

....	Control
—·—·—·—	Control-sand
———◆———	Nanoparticle-beads
——◆——	Raw Oyster shell
———○———	Pyrolysed Oyster shell
——■——	Raw Crushed coral
———□———	Pyrolyzed Crushed coral
———▲———	Raw Mussel shell
———△———	Pyrolyzed Mussel shell

T_0: time zero (phosphorus source alone)

T_M: time immediately after addition of material

T_S: time immediately after addition of sand

Figure 3.2 Phosphate concentration during the batch experiments with A) phosphate solution, B) phosphate solution and sand, C) domestic wastewater and, D) domestic wastewater and sand. Due to technical problems, the trial with the raw material in graph C, was conducted separately from the rest of the materials, each with its own Control (dotted line, × symbol for the raw material).

Table 3.3 pH ranges in batch experiments using phosphate solution and domestic wastewater.

Material	Phosphate solution	n	Phosphate solution + sand	n	Domestic wastewater	n	Domestic wastewater + sand	n
Control	4.9-5.2	28	5.1-6.1	14	6.9-7.5	33	7.0-7.8	13
Nanoparticle-beads	6.7-7.8	4	7.0-7.5	4	7.1-7.5	4	7.1-7.5	4
Raw oyster-shell	8.7-10.1	4	8.0-8.3	4	7.2-7.6	4	7.1-7.4	4
Pyrolyzed oyster-shell	12.5-12.8	4	12.5-12.6	4	12.5-12.6	4	12.4-12.6	4
Raw crushed coral	7.4-7.8	4	7.6-7.8	4	7.2-7.8	4	7.1-7.4	4
Pyrolyzed crushed coral	12.3-12.5	4	12.3-12.5	4	8.9-12.3	4	12.5-12.6	4
Raw mussel shell	7.4-8.1	4	7.4-7.9	4	7.1-7.6	4	7.1-7.4	4
Pyrolyzed mussel shell	12.5-12.8	4	8.8-12.6	4	12.4-12.6	4	9.8-12.6	4

Adding NaOH to the bottles containing raw material and phosphate solution, to raise the pH to approximately 12 (similar to that reached when using pyrolyzed material, Table 3.3), dramatically increased the immediate removal of phosphorus ("$T_{M\ \&\ pH\ adjusted}$") from 22 mg L^{-1} to approximately 13 mg L^{-1} (Fig. 3.3). The subsequent results (1 h and 1 d) showed further phosphorus removal only for the high-pH raw material and at 15 d, they reached phosphorus removal efficiencies similar to those achieved by the pyrolyzed material.

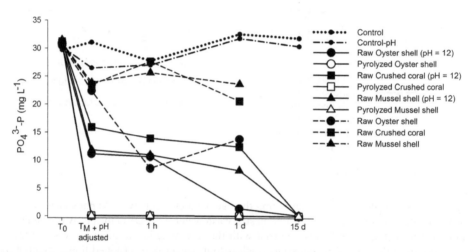

Figure 3.3 Phosphate concentration in the high pH raw material experiment (solid lines). Dashed lines belong to a separate experiment shown in Fig. 3.2A, but were added in this graph for easier comparison. Abbreviations: T_0, time zero (phosphorus source alone); $T_{M\ \&\ pH\ adjusted}$, time immediately after the material addition and adjustment of pH to 12.

3.3.1.2. Engineered material

The immediate phosphorus removal of NB was very small (< 6%) with both phosphorus sources, but it increased to 64% (phosphate solution) and to 25% (wastewater) when sand was present, although no phosphorus was removed in the control-sand at T_S and T_M (Fig.

3.2). During the first hour, almost no phosphorus removal occurred without sand (12-21%), but with sand the removal was higher and reached values of 57-76%. Within 7 d, all experiments reached values as those reached by the pyrolyzed material (> 99% removal). When the phosphate solution came in contact with the NB (both with and without sand), the pH was raised to neutrality, while in the case of the wastewater, the pH was maintained at neutral values.

3.3.2. Column experiments

During the first 30 d of raw wastewater application, the same trend as encountered in the batch experiments was found: pyrolyzed material and NB (with and without sand) removed phosphorus with a higher efficiency (above 99%) than that achieved by the raw materials. The columns containing raw OS and sand, on the contrary, progressively released phosphorus and in many cases the effluent phosphorus concentration was higher than that of the influent. The performance of the pyrolyzed OS and NB columns (with and without sand) deteriorated after a month of operation, increasing the effluent phosphorus concentration to close of that of the influent levels at day 61, when the column's operation was terminated (Fig. 3.4).

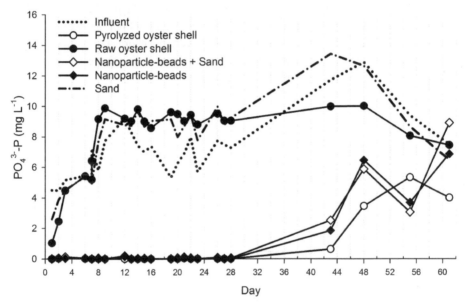

Figure 3.4 Phosphate concentration in the influent and effluent of the columns used in this study (flow rate 0.5 mL min^{-1}). Effluent results of all columns (except sand) are the average of duplicate values.

The pH of the influent and all effluents, except of the pyrolyzed OSs, were similar to the values found during the batch experiments and ranged from 7.0 to 8.4, while the effluent pH of the pyrolyzed OSs columns ranged between 11.7 and 12.7. The Ca^{2+} concentration in the effluent of the raw and pyrolyzed OS columns was 82 ± 17 and 431 ± 181 mg L^{-1}, respectively, much higher than that in the influent (65 mg L^{-1}, Table 3.1).

3.3.3. Phosphorus removal mechanism of the pyrolyzed material

The immediate phosphorus removal (and high pH values) encountered when adding the pyrolyzed material in the batch experiments suggested that the phosphorus removal mechanism used by the pyrolyzed material was precipitation. The Imhoff cone test showed a slight increase of solids from 16 to 19 mL L^{-1} when water and phosphate solution were used, respectively. It was visible that the texture of the extra 3 mL L^{-1} solids was different (e.g. little flocs, light material) than that of the pyrolyzed material granules. The leaching test revealed that no release of phosphorus from the sample was observed after 24 h of mixing in the leaching device, showing a similar phosphorus concentration as in the control. Results of testing the samples with and without solids and colloids revealed that the phosphorus was completely removed in both cases. The addition of acid in the titration experiment disturbed the removal of phosphorus (Fig. 3.5A); instead of above 99% removal (solid line) a maximum of only 33% was reached when the pH was kept in neutral conditions (dashed line). Figure 3.5B shows the equilibrium solubility diagram of hydroxyapatite precipitates where the pH is plotted against the phosphorus concentration. At pH 12, the phosphorus concentration is about 0 mg L^{-1}, indicating that the precipitates did not dissolve. At a pH of approximately 8, a little increase in solubility is observed (phosphorus concentration of 1.76 mg L^{-1}). Below pH 8, the phosphorus concentration increased to 15-30 mg L^{-1}, equal to the initial phosphorus concentration in the phosphate solution.

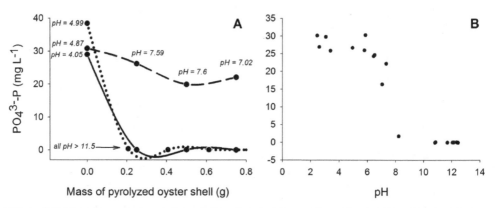

Figure 3.5 Phosphate concentration during the titration with acid test (A) and phosphate equilibrium solubility diagram (B). In graph A, the dashed and solid lines show the results in the presence and absence of acid, respectively. The dotted line shows the results of the titration with CaO (without acid).

3.4. DISCUSSION

3.4.1. Pyrolyzed marine materials

The pyrolyzed marine materials were the most efficient phosphorus removal materials during the batch experiments. Their contact with the phosphorus source immediately decreased the PO_4^{3-}-P concentration to undetectable levels. This is in agreement with the findings of Lee et al. (2009). However, the amount of pyrolyzed material used in the batch experiments was an overdose of > 80 times according to titration results (Fig. 3.5A, solid line) and of > 158 times, to the stoichiometry equation (Eq. 3.1). Hence, less pyrolyzed material would also have immediately removed the same amount of phosphorus. Lee et al. (2009) suggested a final dosage of 0.017 g pyrolyzed OS g^{-1} of phosphorus.

$$10\ Ca^{2+} + 6\ PO_4^{3-} + 2\ OH^- \leftrightarrow Ca_{10}(PO_4)_6(OH)_2 \hspace{3cm} \text{Eq. 3.1}$$

The Imhoff cone test implied that precipitation is one phosphorus removal mechanism and the other experiments (i.e. the leaching test, the experiment with/without solids and colloids and the titration with/without acid) confirmed that precipitation alone provides (almost) all the phosphorus removal rather than adsorption. For adsorption, a surface that provides the adsorption must be present. Yet, when all the solids and colloids were removed in the "solids/colloids experiment", phosphorus removal still occurred, indicating that precipitation was the main removal mechanism. Studies by Abeynaike et al. (2011) and Lee et al. (2009) showed similar results for pyrolyzed MSs and pyrolyzed OSs, respectively. Furthermore, Abeynaike et al. (2011) describe two types of precipitation: (i) homogeneous, which occurs in the reaction solution and (ii) heterogeneous, which occurs on the pyrolyzed shell surface. In the "solids/colloids experiment", when the solids and colloids were filtered out from the solution, the phosphorus removal was as high as it was in the presence of solids and colloids, suggesting that the reaction takes place in the solution. Therefore, homogeneous precipitation is likely occurring.

OSs are composed of 96% $CaCO3$ (Yoon et al., 2003) and the action of pyrolysis converts the insoluble $CaCO_3$ to CaO according to:

$$CaCO_3 + heat \leftrightarrow CaO + CO_2 \hspace{3cm} \text{Eq. 3.2}$$

In our experiments, the presence of CaO was supported by the almost identical results obtained during the titration of the pyrolyzed material and of the pure CaO (Fig. 3.5A, solid and dotted line, respectively). CaO in contact with water dissociates into calcium and hydroxyl ions (Eq. 3.3), raising the solution pH (Table 3.3). The phosphorus in the solution reacts with both ions to form calcium phosphates that will precipitate as hydroxyapatite in the solution according to Equation 3.1.

$$CaO + H_2O \leftrightarrow Ca(OH)_2 \leftrightarrow Ca^{2+} + 2OH^- \hspace{3cm} \text{Eq. 3.3}$$

It should be noted that, since the initial pH of the phosphorus sources was different, the main forms of phosphate present were different as well: $H_2PO_4^-$ in both sources and HPO_4^{2-} only in the domestic wastewater. However, it should be clear that, the initial speciation differences between the solutions did not influence the obtained results. In both the drastic increase of pH (to approximately 12) achieved by the use of pyrolyzed material, rapidly converted any phosphorus form to HPO_4^{2-} and PO_4^{3-}. The latter is involved in the formation of hydroxyapatite (Eq. 3.1).

3.4.2. Raw marine materials

The raw materials tested in this study did not show a high level of immediate phosphorus removal efficiency (Figs. 3.2 and 3.4), suggesting that precipitation did not play an important role (pH < 8) (Abeynaike et al., 2011). Park and Polprasert (2008) showed that the phosphorus removal mechanism by raw OS was adsorption. Only when the pH was manually increased to approximately 12, the phosphorus removal doubled (Fig. 3.3) due to precipitation induced by the Ca^{2+} ions from the $CaCO_3$ of the raw material and by the high pH. However, $CaCO_3$ is less soluble than CaO, thus there were probably not enough Ca^{2+} ions to induce precipitation and further increase the phosphorus removal efficiency of the raw marine materials. Indirect confirmation is provided by the Ca^{2+} effluent concentration measured; where the pyrolyzed OS columns (431 ± 181 mg L^{-1}) discharged higher Ca^{2+} concentration than the raw OS columns (82 ± 17 mg L^{-1}).

In the batch experiments, the raw material achieved lower removal efficiencies as compared to the pyrolyzed material (Figs. 3.2 and 3.4) but higher than what many conventional CWs (sand/gravel beds) remove. Some authors have used raw marine material in laboratory setting CWs to enhance the phosphorus removal of CWs to > 88% and phosphorus effluent concentrations of > 1.3 mg L^{-1} (Wang et al., 2013; Park and Polprasert, 2008). In the column experiment, the effluent concentration of the raw marine material and the sand columns was higher than that in the influent (Fig. 3.4). This can be attributed to desorption occurring after saturation of the particles and the phosphorus concentration in the particles increasing beyond that of the pore water (Dunne and Reddy, 2005).

3.4.3. Engineered material

In all batch experiments, the NBs achieved removal efficiency similar to that of the pyrolyzed material but only after 7 d (Fig. 3.2) and pH values in the range of 6.7-7.8 (Table 3.2). However, a similar batch experiment was conducted (with the phosphate solution) but this time adjusting the initial solution pH (T_0) to neutral values (Appendix A). In this experiment, the phosphate removal efficiency of 10 g of NBs was above 99% within 3 h and the pH was > 8.5 within 1 h (Appendix A). This shows the direct link between the pH and phosphate removal by the NBs.

Around a neutral pH, $H_2PO_4^-$ and HPO_4^{2-} are present. Depending on the phosphate form, the complex can form due to monodentate ($H_2PO_4^-$) or bidentate (HPO_4^{2-}) bonds; the latter being the stronger bond. Above pH 8.5, HPO_4^{2-} predominates and tends to form bidentate complexes, thus enhancing the phosphate removal (Blaney et al., 2007). With domestic wastewater (pH of 6.9-7.5, Table 3.2) the phosphate removal took longer probably also due to solids accumulation and competing ions (e.g. SO_4^{2-}, Cl^-) (Blaney et al., 2007). The NBs used in this study were anion exchangers (positively charged) that adsorb the negatively charged compounds via Coulombic interaction (or ion exchange capacity - forming outer sphere complexes with any negatively charged molecule) and Lewis acid base interaction (forming inner sphere complexes - which are more stable than outer sphere complexes - exclusively with phosphates) (Blaney et al., 2007).

According to calculations, 1 g of NBs can remove approximately 1.35 mg of P (0.74 mg P g^{-1}, Appendix A). The NBs capacity is relatively low when comparing with other man-made products listed in Vohla et al. (2011) or with the list of phosphorus adsorbents summarized in Nur et al. (2014). For instance, the ion exchange resin studied in Nur et al. (2014) showed an adsorption capacity of 48 mg P g^{-1}.

During column experiments, the NBs performance was identical to that of the pyrolyzed OSs. This might be due to the different material-to-solution ratio (1:50 for batch experiments and 1:0.5 for column experiments) and the amount of sorption sites exposed to the solution (Cucarella and Renman, 2009). In the columns, the combination of downflow, small water volume, and excess of sorption sites induced an expedient phosphorus removal despite the short bed contact time of approximately 1.2 h (measured porosity: 50%).

3.4.4. Effect of sand

Sand was used in the batch experiments to study the phosphorus removal capacity of the materials mixed within the CW media. The mixtures of sand with raw material and with NBs had a marked positive effect on the removal capacity of phosphorus (Fig. 3.2). With the pyrolyzed material no effect was observed, because all the phosphorus was immediately removed by the precipitation process. Since the sand used was composed of low quantities of Fe_2O_3, Al_2O_3, CaO and MgO (0.03-1.4%, according to manufacturer data), the removal of phosphorus due to sorption (adsorption and precipitation) reactions with Mg, Fe, Al and Ca ions (Arias et al., 2001) might have slightly improved phosphorus removal. Moreover, it is feasible that the sand is adsorbing competing ions (e.g. SO_4^{2-} present in the domestic wastewater), slightly increasing the amount of available adsorption sites of the NBs and raw marine material (Fig. 3.2D vs. C).

It should be noted that, when using phosphate solution, the control-sand almost did not remove phosphorus while the mixture of NBs and sand (Fig. 3.2B) removed it faster than in the absence of sand (Fig. 3.2A). This showed that the role of sand in the sorption was low, but not because of other ions competing with phosphorus for the sites, since the phosphate solution does not contain competing ions. Thus, further studies are needed to understand and optimize the phosphorus removal using NBs/sand CW beds.

3.4.5. Engineered vs. pyrolyzed marine material: comparison and practical use

Due to the low performance of the raw marine material as compared to the pyrolyzed and engineered material, it was not proposed for further application and thus, not included in this section. The application of NBs is simple as they did not induce a significant change in water's pH, unlike the pyrolyzed marine material (Table 3.3). For the latter, extra steps must be taken to reduce the pH before disposal. For instance, using a carbonation process (dissolution of CO_2 in the water) followed by clarification to remove $CaCO_{3(s)}$ that forms in this process (Jenkins and Hermanowicz, 1991). Nevertheless, the high pH (~12) produced by the pyrolyzed material can be beneficial for pathogens and virus elimination (Sproul, 1980; Parmar et al., 2001). On the other hand, the use of marine materials for phosphorus removal can reduce the waste from shells generated in many coastal areas and aquaculture sites.

Some practical recommendations can be derived from this study. With a phosphorus concentration of 3 mg L^{-1} and a flowrate of 100 m^3 d^{-1} (~1000 PE, 300 g P d^{-1}), stoichiometric calculations show that approximately 1.3 kg of pyrolyzed OS beads are needed daily. In practice, it will be better to add more beads to avoid daily replacement and because some Ca^{2+} is lost with the effluent. Also filter design should provide enough space for the precipitates as well as allow movement of the beads for better contact with the water and release of Ca^{2+} ions. The preparation of beads is critical to the filter's success. If used in powder form, cementation of the material can occur in the filter, and it completely clog. Lee et al. (2009) pointed to the benefit of easy handling of granular material (1-1.4 mm) rather than fine powders. Sizes of 1-1.4 mm may still cause cementation if the material is packed too tightly in a filter. Since phosphorus removal occurs immediately, the contact time is not a major design parameter; therefore, short retention times can be applied.

In the case of NBs, the contact time is a major issue, as the removal mechanism is adsorption, and it depends on the water pH and material-to-solution ratio. Approximately 1 g of NBs can remove approximately 1.35 mg of P (Appendix A); thus 222 kg of NBs are needed daily for the above example (300 g P d^{-1}). It should be noted that this value (estimated from the batch experiments using phosphate solution) should be taken with caution as it seems to underestimate the capacity of the NBs. For instance, the NBs mass used in the columns (70 g, Table 3.2) would have been able to remove 94.5 mg P. This would imply that after 11 d the columns would have been exhausted and no more phosphorus could have been removed. However, the columns lasted longer (Fig. 3.4) despite the presence of solids and competing ions.

The NBs are regenerable and the adsorbed phosphorus can be recovered (e.g. Cumbal et al., 2003; Blaney et al., 2007). In contrast, the pyrolyzed material cannot be regenerated, but the precipitates can be recovered and used as fertilizer or in place of lime as a pH adjuster for chemical processes. These differences have high impacts on the costs. While the NBs require a high initial cost, the ability for regeneration reduces the operational costs (manufacturer data). The marine material is an inexpensive source (low initial cost) that must be renewed (after its consumption in the filter) constituting a permanent production cost (pyrolysis and beads production) over and above the regular operational costs. With the available information, it was not possible to provide a detailed cost analysis for the tested materials.

The equilibrium solubility diagram (Fig. 3.5B) provided a key recommendation when using pyrolyzed material: it is extremely important to keep the precipitates inside the column for reuse purposes. They can be released back into the environment if the solution pH drops to neutral levels or lower.

In the pyrolyzed OS columns, the total amount of material used resulted in a life span of approximately 167 d. Calculations were made using Equation 3.1 and the Ca^{2+} influent and effluent concentrations. They were based on the assumptions that pyrolyzed material is 100% CaO and that homogenous water distribution exists in the column. Solids that could clog the column were disregarded. However, all the columns operated properly for only one month (Fig. 3.4) due to clogging (the raw wastewater provided approximately 98 mg TSS d^{-1}). The solids deposited around the material particles impeded the release of Ca^{2+} ions from the pyrolyzed material to the wastewater (to enhance precipitation) and the contact of the wastewater with adsorption sites of the NBs. Backwashing was unsuccessful in our columns because the material was fully packed.

This study showed that the pyrolyzed or NB filter should be designed as a post-treatment rather than as the first step of phosphorus elimination. Using it as post-treatment provides nutrients for the plants in the CW and assures that the solids contained in the wastewater are trapped in the CW medium instead of the phosphorous removal filter. As a security factor, the material should always allow the possibility of backwashing. Therefore, the material should be allowed to expand and the dislodged particles to flow through. Finally, it is important to mention that the use of this post-treatment step is not limited to CWs and could be applied for the removal of phosphorus elsewhere.

3.5. CONCLUSION

Four materials were tested for the removal of phosphorus from wastewater to be applied as a post-treatment for CWs, leading to the following conclusions:

▪ The pyrolyzed marine material and the NBs provided an efficient phosphorus removal exceeding 99%, and are suitable for a CW post-treatment system. In contrast, the raw materials did not provide efficient phosphorus removal.
▪ The major phosphorus removal mechanism of the pyrolyzed marine material is precipitation, thus the importance of an appropriate pyrolyzation process (100% conversion to CaO). Beads should be used in the filter to avoid clogging and the precipitates should be retained to avoid the phosphorus redissolution at neutral pHs.
▪ The high pH (~12) of the effluent after the treatment with pyrolyzed material can be a limitation if it is intended to be disposed in water bodies and should be corrected. For other purposes, the high pH can be beneficial (e.g. disinfection).
▪ The adsorption capacity of the NBs is determined by the solution pH. Solution pH

above neutral values enhances the removal of phosphorus by NBs.

▪ Installing a separate post-CW treatment system for phosphorus removal improves the performance of a CW without having to design a CW larger system to provide adequate phosphorus removal.

CHAPTER 4.

EFFECT OF AERATION ON POLLUTANTS REMOVAL, BIOFILM ACTIVITY AND PROTOZOAN ABUNDANCE IN CONVENTIONAL AND HYBRID HORIZONTAL SUBSURFACE-FLOW CONSTRUCTED WETLANDS

The large area demand of constructed wetlands (CWs) is documented as a weak point that potentially can be reduced by applying active aeration. The aim of the study was, therefore, to understand the effects of aeration on the treatment performance, the biofilm activity, the protozoan population size and potential CW footprint reduction of different horizontal flow CW configurations. Two experimental periods were considered that tested different organic loading rates (OLR): a first period with 11 g COD m^{-2} d^{-1} and a second period with 20 g COD m^{-2} d^{-1}. Three horizontal flow CW configurations were compared: a conventional (Control), an Aerated and a Hybrid CW (aerated followed by a non-aerated CW). The obtained results reinforced the competence of Aerated CW for organic matter removal (81-89% of chemical oxygen demand) while for nitrogen elimination the Control (19-24%) and Hybrid (8-41%) systems performed better than the Aerated system (-6-33%). Biofilm activity and protozoa abundance were distinctly higher at the inlet zones when compared to the outlet zones of all CWs, as well as in the aerated systems as compared to the non-aerated CWs. The protozoan abundance increased with an increase in the OLR and ciliates were found to be the dominant group. Overall, the active aeration highlighted the efficiency and stability of the CWs for organic matter removal and thus can be used as a promising tool to enhance microbial activity and grazing by protozoa; eventually reducing solid accumulation in the bed media. These beneficial effects contribute to reduce the CWs' area requirements.

This chapter is based on:
Zapater-Pereyra M., Gashugi E., Rousseau D.P.L., Alam M.R., Bayansan T., Lens P.N.L. (2014), "Effect of aeration on pollutants removal, biofilm activity and protozoan abundance in conventional and hybrid horizontal subsurface-flow constructed wetlands", Environmental Technology, 35 (16), 2086-2094.

4.1. INTRODUCTION

The natural, low-cost and aesthetic attributes, coupled to the high contaminant removal efficiency of constructed wetlands (CWs) have positioned them as one of the preferred (decentralized) wastewater treatment systems in many places, especially for the treatment of domestic wastewater (García et al., 2010). Nevertheless, CWs require large land areas for providing adequate wastewater treatment to meet effluent discharge standards. This large footprint is usually a disadvantage, especially when land is scarce due to economic or geographic reasons.

To reduce the surface area required by a CW and maintain the same treatment performance, a few studies have used a design that stacks different treatment systems (e.g. Troesch et al., 2010), but most have tried to raise the systems efficiency per unit area. Common practices for increasing the efficiency are bed oxygenation and wastewater recirculation. The presence of oxygen within the bed is a crucial factor for increasing both biological activity and stimulating nitrification. The main advantage of wastewater recirculation is to increase the contact time of the wastewater with the prevailing biofilm, while it at the same time increases the transfer of oxygen to the wastewater aiding with bed oxygenation (Sun et al., 2003).

The first attempt to increase the oxygen availability within the bed was when vertical flow CWs were developed, reducing the required area of 5 m^2 PE^{-1} (population equivalent) for horizontal flow (HF) to 1-3 m^2 PE^{-1} (Vymazal, 2011). Later, various ways of enhancing aeration were tested, among others, shallow depth of the CW (García et al., 2005), tidal flow systems (Sun et al., 2005), passive oxygenation either by allowing the wastewater drops to be in contact with the atmosphere while falling from an upper to a lower stage (Sklarz et al., 2009; Ye and Li, 2009) or by embedding pipes along the bed with one end opened to the atmosphere and the other connected to the drainage pipes (Brix and Arias, 2005) and artificial aeration by adding pipes within the bed connected to an air blower (Wallace, 2001; Nivala et al., 2007).

From the above mentioned endeavors, increasing attention has been given to artificial aeration as it has demonstrated benefits when removing organic matter, suspended solids and nutrients, either in surface flow CWs (Jamieson et al., 2003), HF CWs (Wallace, 2001; Ouellet-Plamondon et al., 2006; Nivala et al., 2007; Chazarenc et al., 2009; Zhang et al., 2010) or vertical flow CWs (Tang et al., 2009). Nivala et al. (2013) highlights that artificial aeration can often show a more than ten-fold increase of removal rates compared to conventional CWs.

Regardless the advantages of artificially aerated CWs, skepticism remains towards its use due to its restriction effects on denitrification and/or because of economic reasons (energy consumption by the air blower). In many cases, the higher costs can be compensated by the beneficial effects in the final effluent quality and clogging mitigation, while the denitrification restrictions can be solved for instance by placing the air diffuser only close to the HF CW inlet (Chazarenc et al., 2009) or using a hybrid system that combines an aerated and a non-aerated component. Therefore, this study centers on understanding the effects of continuous artificial aeration on the performance of a HF and hybrid CW for footprint reduction by comparing two types of organic loading rates when measuring common wastewater parameters, microbial activity and the protozoan population. HF CWs were studied mainly with the intention of combining their anoxic properties with aerobic conditions (artificial aeration) to increase the nitrogen removal rate.

4.2. MATERIALS AND METHODS

4.2.1. Experimental set-up

Three HF CWs (Fig. 4.1) were assembled indoors using plastic containers packed with gravel of 40% porosity (8-16 mm diameter) and planted with common reed *Phragmites australis* grown under artificial light (approximately 40-50 µmol photons m^{-2} sec^{-1} for 16 h d^{-1}).The systems design varied between them: (i) a conventional HF CW used as a Control system, (ii) an Aerated system and (iii) a Hybrid system, composed of a shallow aerated HF CW (Hybrid-A) and a shallow anoxic/anaerobic HF filter (Hybrid-B). The aerated systems included two aeration tubes at the bottom of the container that were connected to an air blower (Fig. 4.1). The Hybrid system was designed to serve two purposes: to reduce the CW footprint and to optimize the nitrogen removal efficiency through sequential nitrification-denitrification. For the former, the intention was to place one system on top of the other (stack design), therefore the depth is enlarged instead of the area; but for practical reasons, both components were placed next to each other and Hybrid-B was kept under air tight conditions, as shown in Figure 4.1. Since the Hybrid-B was not planted it was considered as a horizontal flow filter. The second purpose of the Hybrid system required the combination of an aerated and an anoxic/anaerobic system that included an addition of influent-wastewater to Hybrid-B as an extra carbon source for denitrification.

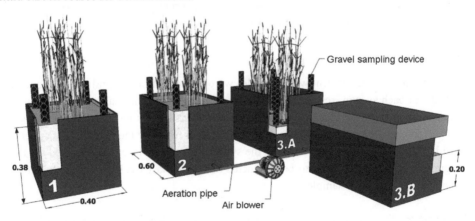

Figure 4.1 Schematic representation of the experimental set-up. The horizontal flow constructed wetlands are: (1) conventional (Control), (2) Aerated and (3) Hybrid. The Hybrid is composed of an aerated component (Hybrid-A, 3.A) and a non-aerated component (Hybrid-B, 3.B). Dimension units are in m. The cuts on the containers surface are there for visibility purposes only.

The surface area of all CWs was 0.24 m^2 (0.6 m × 0.40 m) while the depth of the Control and Aerated system was 0.38 m. Each Hybrid component had a shallow bed depth of 0.20 m, that permitted the same amount of media, and therefore the same surface area for biofilm growth and the same hydraulic residence time (HRT), as the other systems. The water level was maintained at 4 cm below the surface, except for Hybrid-B where the water level reached the gravel surface.

4.2.2. Operating mode

Experiments were conducted for 14 weeks, divided into two experimental periods of five weeks each and four weeks for adaptation. The adaptation took place during the first three weeks (start-up) and during week 9 (between the experimental periods). All systems were fed with primary effluent from a nearby wastewater treatment plant (Harnaschpolder, Delft, The Netherlands) and operated at 2 d HRT, with a continuous flow rate of 16.3 L d^{-1}. The Hybrid-B received two sources of water: effluent from Hybrid-A, and 40% of bypassed influent diverted directly to the system as a carbon source for denitrification.

Two different wastewater strengths were tested in a first (week 4 to 8) and second (week 10 to 14) experimental period. In the first experimental period, the wastewater was diluted to reach an organic loading rate (OLR) of 10.5 g COD m^{-2} d^{-1} (COD, chemical oxygen demand) (11 m^2 PE^{-1}). In the second, the wastewater was used as it was collected (OLR of 19.7 g COD m^{-2} d^{-1}; 6 m^2 PE^{-1}). Based on the calculations of the oxygen transfer rate (using COD and NH$_4^+$-N data; Cooper, 2005) and the air flow rate (Metcalf and Eddy, 2003) for the more concentrated wastewater (second experimental period), any air flow above 3.3 L min^{-1} assured oxygen saturation conditions in all the systems (Appendix B). Hence, the air blower was used to its maximum capacity (15-23.5 L min^{-1}) to provide continuous artificial aeration; but after the first period, severe biofilm loss due to scouring was noticed. Consequently, the air flow rate was reduced to 7.5-11.8 L min^{-1} for the second period.

4.2.3. Sampling and analytical techniques

Sampling. Influent and effluent water samples were collected from the influent tank and from the outlet of each system, respectively. The gravel pieces used for analyses were taken from four sampling devices (Fig. 4.1), each made of two layers of iron mesh cylinder and inserted vertically at each sampling point, two near the inlet and two near the outlet of the systems. It should be noted that, to avoid disturbing the systems, all the analyses that required the use of the gravel media were planned to be conducted during the last weeks of the study, although this meant a reduction in sample size.

Treatment performance. The water samples were immediately analyzed for pH, redox-potential (Eh) and dissolved oxygen (DO) by the electrometric method, COD by the potassium dichromate method, dissolved organic carbon (DOC) by the high-temperature combustion method for total organic carbon (TOC) (Shimadzu TOC-VCPN) after filtering the samples over a prewashed 0.45μm cellulose membrane filter, nitrate and phosphate by the ion chromatography method, total suspended solids (TSS) by the glass-fiber filter method, nitrite by the colorimetric method and total nitrogen (TN) was digested by the persulfate method, followed by NO_3^--N (ultraviolet spectrophotometric screening method), all according to APHA (1998). Ammonium was measured by the dichloroisocyanurate method (NEN 6472, 1983). Organic nitrogen (Org-N) was calculated as the result of the TN minus the sum of ammonia, nitrate and nitrite. Sulfate was measured according to the SulfaVer® 4 (Hach Lange GmbH, Germany) method. All parameters were analyzed seven times during each experimental period (n = 7).

For the determination of the organic matter fractions, the DOC of the sample was further characterized by analyzing the fluorescence excitation(EX)-emission(EM) matrix spectra (Her et al., 2003). Briefly, the water samples were diluted to a final DOC concentration of 1 mg L^{-1} and scanned in a 1.0 cm quartz cell using a computer controlled spectrofluorometer (Horiba Jobin Yvon Fluoro Max-3) equipped with a 150-W ozone free xenon arc-lamp, over an EX range of 240-450 nm and an EM range of 290-500 nm. Three peaks were identified for humic/fulvic-like, humic-like and protein-like compounds, under EX/EM wavelengths of 330-350 nm/420-480 nm, 250-260 nm/380-480 nm and 270-280 nm/320-350 nm, respectively. Data in the form of a matrix were obtained and converted to 3D maps using MATLAB R2007b software. This characterization was conducted at the beginning and at the end of each experimental period (n = 2).

Solids accumulation. Accumulated solids on the gravel samples near the inlet and outlet were measured using the TSS method after 1 h sonication in demineralized water. For that, gravel samples were collected from each sampling device (n = 4 for the Aerated and Control systems and n = 2 for each Hybrid system component) at the end of the study.

Biofilm. The protein concentration in the biofilm was determined by measuring the concentration of Org-N and converting it to protein content based on the assumption that protein contains 16% nitrogen (Boisen et al., 1987). Briefly, gravel samples were gently washed to remove excess ammonia. Then, the gravel samples were sonicated with a known volume of demineralized water for 1 h and 50 mL of this suspension to estimate the Org-N content using the total kjeldahl nitrogen (TKN) method (APHA, 1998).

The fluorescein diacetate (FDA) hydrolysis test was used to determine the microbial activity in the biofilm as described by Iasur-Kruh et al. (2010). Approximately 10 pieces of gravel were taken from each sampling device and gently rinsed with demineralized water. A solution of 0.3 mL FDA (0.05mg mL^{-1}) and 30 mL of buffer stock solution ($Na_2HPO_4.2H_2O$ and KH_2PO_4 at a ratio of 9:1, pH = 7.6) was poured onto the gravel samples and incubated at 25°C for 2 h. The solution was filtered over 0.2 µm and the absorbance of fluorescein was measured at 494 nm. To express the microbial activity as $\mu g_{fluorescein}$ $mg^{-1}_{biomass}$ h^{-1}, the biomass was determined by the volatile suspended solids standard method (APHA, 1998) after sonicating the gravel samples to detach the adhered biomass. The FDA assay was performed once (n = 1) in the last week, while the protein content was determined once a week during the last two experimental weeks (n = 2).

Micro-invertebrate. About 3-4 gravel pieces were taken from each sampling device and the biofilm was scraped off with a small brush. The separated biofilm was mixed with interstitial water and gently shaken. The samples were analyzed within 12 h after their collection with an optical microscope at a magnification of ×20 or ×40, to determine the genus level. A volume of 30 μL was collected with a pipette and three replicates were examined. Several taxonomic keys were employed for the *in vivo* identification of different protozoa (cilliates and flagellates) and one type of metazoa (rotifers) (Streble and Krauter, 2006). The micro-invertebrate population was expressed as population per mL (Puigagut et al., 2007). This analysis was conducted once at the end of each experimental period (n = 1).

4.2.4. Data analysis

One way analysis of variance (ANOVA) followed by the Tukey test for multiple comparisons and Kruskal-Wallis test (H-test) were used, respectively, for parametric and non-parametric distribution to test the differences between the effluent concentrations of the three CWs. The statistical analysis was conducted at a 95% confidence level.

4.3. RESULTS

4.3.1. Treatment performance

4.3.1.1. Organic matter and solids

The effluent COD and DOC concentrations for the Aerated and Hybrid systems were lower than those in the Control system during the second experimental period ($p < 0.05$), while no significant differences between the three effluents were found for COD ($p = 0.75$) and DOC ($p = 0.25$) during the first period (Table 4.1). The overall COD removal efficiency varied between 69-70% for the Control, 81-89% for the Aerated and 72-82% for the Hybrid system, showing that a higher COD removal efficiency was achieved by the Aerated system. The DO and Eh in the effluent slightly dropped in the second period. The average pH of the samples was close to neutrality in the Control and Hybrid systems, whereas it was moderately alkaline in the Aerated system (Table 4.1).

The Aerated and Hybrid systems had low TSS effluent concentrations (Table 4.1), and showed significant differences with the Control ($p < 0.05$) during the first experimental period. In the second period, a high TSS removal efficiency was also noticed (> 89%) in all systems, however the three systems were statistically similar ($p = 0.07$). At the end of the study, more solids had nevertheless accumulated on the gravel near the inlet than at the outlet of each system and near the inlet of the non-aerated systems compared to that of the aerated systems (Fig. 4.2).

Table 4.2 summarizes the peak intensities of humic/fulvic-like, humic-like and protein-like compounds in the wetlands, and one example is visible in Fig. 4.3. The highest reduction in all peak intensities was observed in the Aerated system, followed by the Hybrid system. The highest peak reduction (up to 86%) was observed in the Aerated system for the protein-like compounds.

Table 4.1 Concentrations of the various water quality parameters in the influent and effluent of the constructed wetlands during the two study periods. The values indicate mean ± standard deviations (n = 7). All units are expressed in mg L^{-1} unless otherwise stated.

Parameter	Influent	Control	Aerated	Hybrid
First experimental period - 10.5 g COD m^{-2} d^{-1}				
pH* (unitless)	7.41 ± 0.17	7.33 ± 0.14^a	8.14 ± 0.04^b	7.32 ± 0.09^a
Eh* (mV)	19 ± 155	-195 ± 18^a	48 ± 33^b	81 ± 29^b
DO*	3.7 ± 0.3	0.9 ± 0.2^a	7.7 ± 0.4^b	1.3 ± 0.2^a
COD	154 ± 32	47 ± 14^a	28 ± 9^a	43 ± 20^a
DOC	19.7 ± 6.1	13.3 ± 4.1^a	10.3 ± 2.8^a	11.2 ± 2.8^a
TSS	48.7 ± 9.5	8.3 ± 2.6^a	2.6 ± 0.8^b	3.7 ± 0.9^b
SO$_4^{2-}$	78 ± 10	59 ± 14^a	108 ± 19^b	98 ± 10^b
Org-N	5.4 ± 1.7	5.2 ± 3.2^a	7.3 ± 4.3^a	5.1 ± 3.4^a
NH$_4^+$-N	16.4 ± 3.5	12.7 ± 4.3^a	0.6 ± 0.5^b	3.3 ± 2.4^b
NO$_2^-$-N	0.21 ± 0.19	0.22 ± 0.16^a	0.06 ± 0.05^a	0.73 ± 0.26^b
NO$_3^-$-N	1.67 ± 0.29	0.69 ± 0.42^a	16.76 ± 3.20^b	12.35 ± 0.92^b
TN	23.6 ± 3.7	18.8 ± 4.1^a	24.7 ± 3.2^b	21.3 ± 3.2^a
PO$_4^{3-}$-P	3.18 ± 1.01	3.29 ± 1.33^a	3.12 ± 0.75^a	3.20 ± 0.68^a
Second experimental period - 19.7 g COD m^{-2} d^{-1}				
pH (unitless)	7.28 ± 0.12	7.18 ± 0.06^a	7.87 ± 0.09^b	7.36 ± 0.09^b
Eh (mV)	46 ± 34	-254 ± 24^a	1 ± 16^b	-113 ± 111^b
DO	2.2 ± 0.5	0.5 ± 0.2^a	6.8 ± 0.4^b	0.9 ± 0.3^b
COD	290 ± 50	90 ± 38^a	31 ± 5^b	51 ± 10^a
DOC	27.8 ± 14.4	16.4 ± 5.4^a	9.4 ± 1.8^b	12.6 ± 3.6^a
TSS	109.4 ± 3.7	11.8 ± 3.3^a	5.9 ± 3.4^a	9.5 ± 4.2^a
SO$_4^{2-}$	67 ± 26	23 ± 6^a	65 ± 30^b	53 ± 17^b
Org-N	8.2 ± 4.3	2.3 ± 2.9^a	2.3 ± 1.9^a	3.0 ± 2.6^a
NH$_4^+$-N	25.3 ± 8.3	23.1 ± 7.2^a	0.3 ± 0.3^b	11.0 ± 4.7^b
NO$_2^-$-N	0.16 ± 0.21	0.02 ± 0.02^a	0.04 ± 0.03^a	0.76 ± 1.15^a
NO$_3^-$-N	0.34 ± 0.44	0.63 ± 0.71^a	19.75 ± 6.87^b	5.36 ± 3.94^a
TN	34.0 ± 9.9	26.1 ± 8.1^a	22.5 ± 5.9^a	20.2 ± 8.3^a
PO$_4^{3-}$-P	3.08 ± 1.14	3.44 ± 0.80^a	3.60 ± 0.93^a	3.51 ± 1.01^a

* Due to technical reasons, n = 3.
A different letter (a, b in superscript) for a particular parameter indicates a statistically significant difference with the Control constructed wetland ($p < 0.05$).

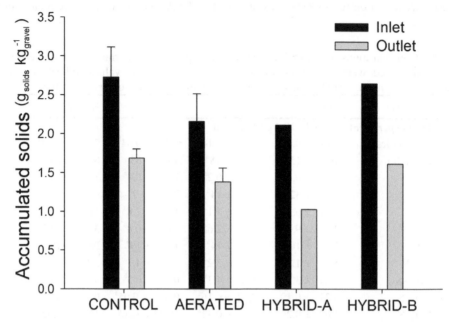

Figure 4.2 Accumulated solids on the gravel surface measured at the end of the study. Error bars are standard errors (n = 4, for the Control and Aerated systems). No error bars were calculated for the Hybrid system columns (n = 2).

Table 4.2 Peak intensities of humic/fulvic-like, humic-like and protein-like compounds (n = 2) in the constructed wetlands.

	Humic/fulvic-like EX/EM (320/424nm)	Humic-like EX/EM (250/436nm)	Protein-like EX/EM (280/326nm)	DOC removal %
First experimental period - 10.5 g COD m^{-2} d^{-1}				
Influent	2.63	4.09	1.68	
Control	1.88	2.75	1.92	38
Aerated	1.52	2.34	0.32	53
Hybrid	1.69	2.59	0.38	50
Second experimental period - 19.7 g COD m^{-2} d^{-1}				
Influent	3.18	4.19	3.19	
Control	3.28	4.39	1.31	33
Aerated	2.03	2.95	0.45	62
Hybrid	2.55	3.60	0.86	49

EX/EM: Excitation/Emission.

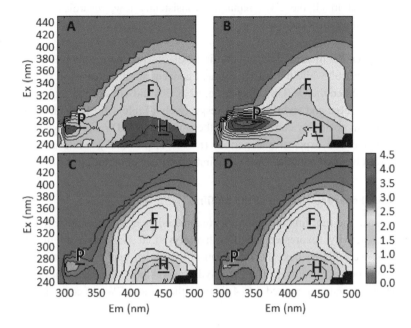

Figure 4.3 Typical Fluorescence Excitation (Ex) -Emission (Em) Matrix spectra during the first experimental period (week 7) for influent (A) and effluent of each constructed wetland: Control (B), Aerated (C) and Hybrid (D). Letters F, H and P stand for the peaks of Humic/Fulvic-like, Humic-like and Protein-like compounds, respectively.

4.3.1.2. Nutrients

The concentration of Org-N in all the systems was characterized by a low removal efficiency during the first period and a high removal efficiency when the higher OLR was applied (second period), ranging between 63-72% removal without significant differences between them ($p = 0.51$) (Table 4.1). The highest reduction in ammonium concentration in both experimental periods was observed in the Aerated and Hybrid systems ($p<0.05$) ranging between 97-99% and 58-81%, respectively, compared to 7-22% for the Control CW. Consequently, nitrate concentrations increased from 0.34-1.67 mg L^{-1} to 16.8-19.8 mg L^{-1} for the Aerated CW and to 5.4-12.4 mg L^{-1} for the Hybrid CW ($p<0.05$). Mean effluent NO_2^--N concentrations increased in the Hybrid system and decreased in the Aerated CW in the two periods, while for the Control system it slightly increased in the first period followed by a decrease in the second period (Table 4.1).

The TN removal in all the systems was relatively low. During the first period, the highest removal efficiency (19%) was observed in the Control system and was statistically similar ($p = 0.37$) to the TN removal (8%) in the Hybrid system (Table 4.1). An improved TN removal of up to 41% was observed in the Hybrid system when the higher OLR was applied (second experimental period), however the three systems were statistically similar ($p = 0.35$). The phosphate removal in all the CWs remained consistently low, regardless of the influent loading and aeration rates (Table 4.1).

The sulfate conversion measured during the first experimental period showed a consumption of 23% in the Control, while the sulfate effluent concentration increased in the Aerated and Hybrid systems. When the OLR was increased, 62%, 5% and 16% of the sulfate was reduced in the Control, Aerated and Hybrid system, respectively (Table 4.1). The sulfate reduction in the Control system positively correlated with the amount of COD removed when operating at the high OLR (second experimental period) ($r = 0.78$, $p < 0.05$), while a non-significant correlation was observed with the low OLR (first experimental period) ($r = -0.53$, $p = 0.22$).

4.3.2. Microbial characterization of the biofilm

Figure 4.4 shows the biofilm concentrations (expressed as mg of proteins per kg of gravel) and microbial activity (expressed as FDA hydrolysis). It was observed that the protein concentrations were higher near the inlet than at the outlet, and lower in the aerated CWs compared to the non-aerated ones. The FDA hydrolysis activity was higher near the inlet than at the outlet and its value increased 3 to 4 times in the aerated CWs as compared to the other systems.

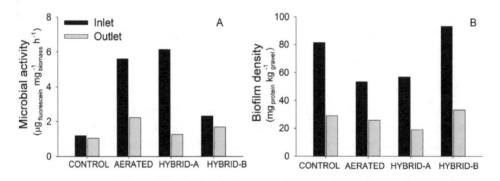

Figure 4.4 Microbial activity ($n = 1$) (A) and biofilm density ($n = 2$) (B) in the constructed wetlands.

Ciliated protozoa were the most abundant protozoa in the three systems investigated (Fig. 4.5). Some other protozoa (i.e. flagellates) and metazoa (i.e. rotifers) were also identified. The free swimming ciliates were the most dominant group among the ciliate species, while others belonged to crawling and stalked ciliates. Concerning the total population at the inlet and outlet of each system, the numbers of protozoa and metazoa in the Aerated system and Hybrid-A were almost 2.5 times higher than in the Control and Hybrid-B systems.

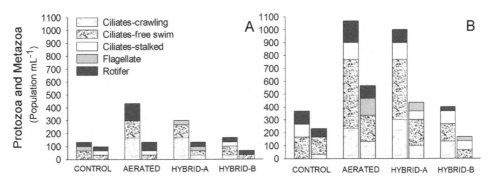

Figure 4.5 Protozoa and metazoa counts of each constructed wetland during the first (10.5 g COD m⁻² d⁻¹) (A) and second (19.7 g COD m⁻² d⁻¹) (B) experimental periods. The two bars per system indicate the population count at the inlet (left bar) and outlet (right bar).

4.4. DISCUSSION

4.4.1. Effect of aeration on organic matter, solids and nutrients removal

This study showed that mechanical aeration significantly enhanced the organic matter removal efficiency, microbial activity and protozoan abundance, but not the nitrogen removal efficiency. It was expected that the Hybrid system would enhance denitrification and thus TN removal, but probably due to a high volume of the bypass influent and/or a short HRT the TN removal was not statistically different from the Control system.

The high SO_4^{2-} reduction and the low DO and Eh results in the Control system confirmed the expected anaerobic/anoxic conditions; thus sulfate reduction was an important pathway for the organic matter degradation in that CW. The high TSS, COD and DOC removal in the aerated systems indicated the importance of aerobic activity to degrade organic matter. The high microbial activity in the aerated systems (Fig. 4.4A) demonstrates the rapid hydrolysis under aerobic conditions and serves as indicator of aerobic biodegradation. The high reduction of the excitation emission matrix peak intensities in the Aerated CW, followed by the Hybrid system (Table 4.2) also emphasizes the competence of aerobic bioconversions in organic matter removal. The reduction in humic/fulvic-like, fulvic-like and protein-like compounds in all systems can be attributed to the significant decrease of aromatic structures throughout the biodegradation processes (Wei et al., 2009). Furthermore, among all organic compounds, protein-like compounds showed the highest reduction, indicating that the easily biodegradable organic matter was converted first, followed by the more complex bonding structures.

The observed significant differences in COD and DOC removal during the second period, opposite to what occurred in the first period, may be related to the high strength of the inlet wastewater. This induced reduced conditions in the systems without aeration, which did not disturb the aerated systems as the aeration rates in both experimental periods were high enough to maintain oxygen saturation (Section 4.2.2). Hence, the presence of continuous artificial aeration in a CW produces more stable conditions in the systems allowing them to withstand OLR fluctuations without altering the in situ working environment.

The average TN removal efficiency observed in the Hybrid system was only 41% (Table 4.1) and, in general, the three systems exceeded the 10-15 mg N L^{-1} of wastewater effluent concentrations permitted in the European Directive 91/271 (EEC, 1991). Nitrification was enhanced in the systems that contained aeration and, therefore, it was considered as the main mechanism of ammonia removal with up to 97-99% removal (Table 4.1), similar to the removal reported by Jamieson et al. (2003) and Nivala et al. (2007). In the Hybrid system, it is possible that denitrification rates were offset by the lack of nitrification of the bypassed influent. Hence, effluent ammonia concentrations increased in Hybrid-B and no conversion to nitrate occurred resulting in the observed incomplete N removal (Table 4.1). Moreover, during the second experimental period (high OLR), the enhanced reduced conditions (DO = 0.9 ± 0.3 mg L^{-1}) limited the nitrification in Hybrid-B, as nitrification is significantly limited at DO < 2 mg L^{-1} (Hammer and Knight, 1994). Furthermore, the HRT of each Hybrid component may have played a role in the low TN removal. The nitrification rate is much slower than the denitrification rate (Verhoeven and Meuleman, 1999), and perhaps the HRT was too short to allow full nitrification and subsequent denitrification to completely remove all N (Table 4.1).

4.4.2. Effect of aeration on microbial community interactions

Oxygen supply increases bacterial counts and microbial activity (Chazarenc et al., 2009) as shown in Figure 4.4. Moreover, it stimulates protozoan abundance (Fig. 4.5) due to their sensitivity to low DO concentrations that usually limit their role when anaerobic conditions predominate (Sinclair et al., 1993). A similar effect is caused by high organic matter concentrations (Decamp et al., 1999; Puigagut et al., 2007) as encountered during the second experimental period and in the untreated wastewater at the CW inlet (Figs. 4.4 and 4.5). In addition, large protozoa populations increase the predation rate of bacteria and consequently enhance the bacterial activity (Hahn and Höfle, 2001; Figs. 4.4 and 4.5). However, the predation rate was still lower than the bacteria growth rate (Figs. 4.4 and 4.5; Matz and Kjelleberg, 2005).

The low biofilm concentration in the aerated systems (Fig. 4.4B) could be the result of biofilm detachment due to abrasion caused by the mechanical aeration (Wuertz et al., 2003) and the protozoa population causing biofilm sloughing through their movement (Huws et al., 2005) and predation (Characklis and Wilderer, 1989). This also yielded a lower solid accumulation in the aerated systems as compared to the non-aerated systems, coupled with the higher microbial activity in those systems that enhanced solids mineralization (Fig. 4.2). In contrast, the larger biofilm concentration at the inlet of the non-aerated systems as compared with the aerated systems (Fig. 4.4B) might be due to the combination of the lack of mechanical aeration and the small protozoa population present (Fig. 4.5).

4.4.3. **Footprint of constructed wetlands**

An area calculations using the first order kinetic equation (assuming no background concentration, $C* = 0$ mg L^{-1}, Appendix C) suggested that, for COD removal, the Aerated and Hybrid systems required 1.9 and 1.5 less area, respectively, than that needed by the Control system during the second experimental period. For NH_4^+-N removal, the areas could be reduced up to 49 and 13 times, respectively. Despite the Hybrid system performance did not seem promising when compared with the conventional HF CW (Control), as both systems were in most of the cases statistically similar (Table 4.1), it still halves the area needed by the Control system to achieve the same TN effluent concentration. Increasing the medium depth (and thus the HRT) and/or decreasing the amount of bypass wastewater are key points to boost the TN and organic matter removal efficiency of CWs treating domestic wastewater and to reduce the area requirements. Thus, different modifications done to the Hybrid design are studied in Chapter 5 and 6.

4.5. CONCLUSIONS

▪ The use of active aeration in HF CWs demonstrated that it can improve the system performance when there is an excess of organic matter and ammonia in the wastewater.

▪ Solids accumulation in the CW media can be controlled by means of enhancing biofilm detachment through mechanical (abrasion) and biological (bacteria grazing by protozoa) processes, potentially providing lesser risk of clogging and a longer life span of the CW.

▪ Artificial aeration can increase the systems efficiency per unit area, and subsequently reduce the HF CW footprint.

CHAPTER 5.

AERATION AND RECIRCULATION IN A STACK ARRANGED HYBRID CONSTRUCTED WETLAND FOR TREATMENT OF PRIMARY DOMESTIC WASTEWATER

The Duplex-constructed wetland (CW) is a hybrid system composed of a vertical flow (VF) CW on top of a horizontal flow filter (HFF). Each compartment is designed to play a different role: aerobic treatment in the VF CW due to intermittent feeding and anoxic treatment in the HFF due to saturated conditions. Three Duplex-CWs were used in this study: Control, Aerated and Recirculating. The Control had a 1 d hydraulic retention time (HRT) in the VF CW and 3-4 d HRT in the HFF while the others operated similarly to the Control system but including aeration in the VF CW (air flow 5 L min^{-1}) and recirculation from the HFF back to the VF CW (10 L h^{-1} for 17 h), respectively. The role of each compartment was tested for pollutant removal (e.g. nitrogen and pathogens), microbial activity, protozoa abundance and potential nitrification and denitrification rates. In all systems, the VF CW removed mainly organic matter, solids and NH_4^+-N. Pathogens were removed in both compartments. Likewise, TN removal occurred in both compartments depending on the system, except for the Recirculating HFF that was not able to denitrify the nitrogen due to the slightly more oxic conditions as compared to the other systems. Thus, all systems met discharge guidelines for organic matter, but only the Control and Aerated systems met those for TN. The pollutant removal was not significantly enhanced by the use of aeration and recirculation. Therefore, at the applied loading rates, operation as in the Control system is recommended.

This chapter is adapted from:
Zapater-Pereyra M., Kyomukama E., Namakula V., Bruggen van J.J.A., Lens P.N.L., "The effect of aeration and recirculation on a sand based hybrid constructed wetland treating low strength domestic wastewater", Submitted to Environmental Technology.

5.1. INTRODUCTION

The raising population in the world brings dire consequences to the water sector such as the increment of fresh water demand and the subsequent decrease of fresh water sources, the massive wastewater production that requires more sanitation technologies and the dwindling of (green) areas due to extensive construction. Constructed wetlands (CWs) are recognized as efficient wastewater treatment systems and appealing green spaces able to alleviate many of the water sector demands. However, the large space required by constructed wetlands (CWs) is usually a disadvantage especially when land availability is scarce. Potential endeavors to obtain a compact and efficient CW include the use of aeration, recirculation or a design that stacks different treatment stages above each other instead of next to each other (Zapater-Pereyra et al., 2014 - Chapter 4).

A potential system to reduce the space requirement is called Duplex-CW (Zapater-Pereyra et al., 2015a - Chapter 6). The Duplex-CW is a hybrid system that consists of a vertical flow (VF) CW placed above a horizontal flow filter (HFF) and that was designed to provide a complete treatment by combining aerobic (the VF CW) and anoxic (the HFF) compartments, without increasing the total footprint of the system (Zapater-Pereyra et al., 2015a). Little is known about the Duplex-CW design. Therefore, this study concentrates on the effect of aeration and recirculation on each compartment (VF CW and HFF) of the Duplex-CW for the treatment of domestic wastewater with special attention to microbial activity and micro-invertebrate (e.g. protozoa and metazoa) abundance. Furthermore, the current and potential treatment capabilities for nitrogen removal were analyzed to understand the current design and future improvements that can be done to reach the maximum treatment capacity.

5.2. MATERIALS AND METHODS

5.2.1. Experimental setup

Three laboratory scale Duplex-CW systems, planted with *Phragmites australis*, were constructed in a greenhouse at the UNESCO-IHE Institute for Water Education (Delft, The Netherlands): a Control, an Aerated and a Recirculating system. The greenhouse was kept approximately at a temperature of 20-23°C and the light intensity was set to 85-100 μmol photons m^{-2} sec^{-1} for 16 h d^{-1}. The compartments (VF CWs and HFFs) were arranged as shown in Figure 5.1. The support medium was coarse sand (1-2 mm) and the drainage layer consisted of gravel (15-30 mm). Each Duplex-CW had a surface area of 0.24 m^2, while the depths were 0.80 m (0.70 m of sand and 0.10 m of drainage layer) for the VF CW and 0.35 m (only sand) for the HFF (Fig. 5.1, Appendix E). Domestic primary wastewater, collected from the Harnaschpolder wastewater treatment plant (Delft, The Netherlands), was allowed to settle for ~2 h and the supernatant was fed to each VF CW compartment (3 times a day, 13 L each time, twice a week). All systems received a hydraulic loading rate (HLR) of ~0.05 m^3 m^{-2} d^{-1} and an organic loading rate (OLR) of 13 g COD m^{-2} d^{-1}, which corresponded to an area of 8 m^2 PE^{-1}.

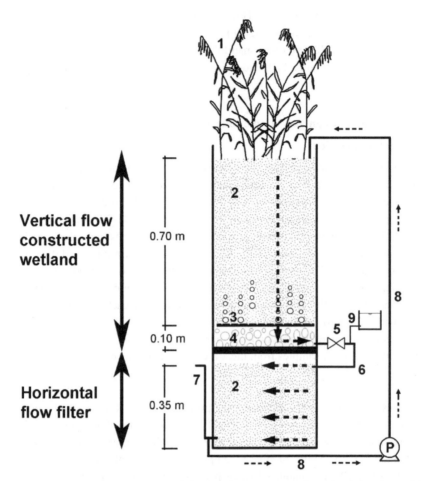

Vertical flow constructed wetland

0.70 m

0.10 m

Horizontal flow filter

0.35 m

Figure 5.1 Schematic representation of the Duplex-constructed wetland (CW) used in this study. 1- *Phragmites australis*, 2- Sand (support media), 3- Aeration pipe, 4- Gravel (drainage layer), 5- Valve, 6- Pipe connecting the compartments, 7- Outlet pipe, 8- Recirculation system and 9- Bucket and pipe to deliver the filtered influent as electron donor for denitrification in the horizontal flow filter. The dashed lines show the path of the wastewater in the system. The connections 3 and 8 were only used in the Aerated and Recirculating Duplex-CWs, respectively.

The three systems operated in a "fill and drain" mode as described in Zapater-Pereyra et al. (2015a) (Fig. 6.2A, Chapter 6). In the Control system, the wastewater stayed for 1 d hydraulic retention time (HRT) in the VF CW, afterwards the outlet valve was opened and the wastewater was released into the HFF, where it stayed 3-4 d. The Aerated system was operated identical to the Control system, but in addition air was supplied to the VF CW during the 1 d HRT at a rate of 5 L min^{-1} (~0.3 m^3 h^{-1}) (Appendix B) through a perforated plastic tubing connected to an air flow pump and meter (Fig. 5.1, Appendix E). In the

Recirculation setup, the water from the HFF was pumped back to the VF CW at a rate of 10 L h^{-1} from where it trickled down into the HFF. This was done by means of a pump connecting the effluent pipe of the HFF to a pipe manifold placed on top of the VF CW (Fig. 5.1). The recirculation was stopped after 17 h and the water stayed in the HFF until the next feeding cycle.

5.2.2. **Experimental design and sample collection**

5.2.2.1. Regular operation

After 3 weeks of adaptation period, the systems were studied for 5 weeks. Water samples were taken weekly from each of the three setups (n = 5). Each time, a total of 9 water samples were collected: influent, 3 positions along the vertical sand profile of the VF CW (from bottom to top: 0-0.20 m (bottom), 0.20-0.45 m (middle) and 0.45-0.70 m (top)), and effluent of the VF CW (= influent of the HFF or "partial effluent"), 3 positions along the horizontal profile of the HFF (from inlet to outlet: 0.05 m (inlet), 0.2 m (middle) and 0.45 m (outlet)) and effluent of the HFF. The maximum water level in the VF CW reached ~ 39 cm, but since the water application occurred at the top, it was possible to sample (interstitial) water at the top layer. Sampling of the influent was done in the tank at each feeding moment. Water sampling within the vertical and horizontal profile was done by gently sucking the water using a syringe attached to a cylindrical core. The effluents were taken directly from sampling ports. All water samples from the VF CWs and HFFs were taken at the end of their HRT.

Sand samples were taken four times along the profile of the VF CWs (mixed with few plant roots) and HFFs, as for water samples. The HFF was only opened for collecting soil samples and covered immediately to avoid oxygen from entering this compartment. Different analyses were conducted with some of these samples: (i) potential nitrification (PNR) and denitrification rate (PDR), conducted twice per sampling point, but since in this study the results per compartment are shown (Table 5.2), then the sample size (n) was 6 for each test, (ii) protozoa and metazoa determination (n = 3) and (iii) microbial activity tests, conducted for the sand (n = 4) and roots separately. Since roots were not always obtained when sampling, the sample size varied and therefore only the mean (± Std. error) results of all depths are presented (n for the Control, Aerated and Recirculating systems are 6, 7 and 7, respectively). The comparison of the activity of very light materials (roots) with heavy materials (sand) should be done in terms of volume (per mL) instead of mass (per g). In this study it was not possible to obtain the weight to volume ratio of the roots, thus a conversion factor of 0.083 g mL^{-1} from the Duplex-CW setups studied in Chapter 6, running for 435 d, was used.

5.2.2.2. **Operation with the addition of a carbon source to the HFF to stimulate denitrification**

Results obtained during the regular operation indicated that the Recirculating setup was not capable to denitrify the available nitrogen (effluent NO_3^--N > 30 mg L^{-1}). A carbon source (as electron donor) was applied to the HFF (Fig. 5.1) to investigate its effect on denitrification. Approximately 3.5 L of *filtered influent* was applied to the HFF during a period of 3 weeks (1 week of adaptation period + 2 weeks of experiment). The type (*filtered influent*) and volume (3.5 L) of the water applied were obtained from preliminary PDR batch test and stoichiometry calculations, respectively, given in Appendix D. The filtered influent used as carbon source refers to the wastewater used to feed the VF CWs after filtration with a piece of cloth. During the regular operation, the VF CWs received 39 L in a feeding day and after 1 d it was transferred to the HFFs. In this experiment, to keep the water volume received by the HFF constant, the influent volume applied to the VF CW was 35.5 L (39 - 3.5 = 35.5 L, 2 batches of ~13 L and 1 batch of 9.5 L). The filtered influent was applied at the moment the VF CW was drained, for mixing purposes. This experiment was also conducted in the Control system for comparison. Water (n = 3) and sand (n = 3) samples were taken from the HFF at the same points as in Section 5.2.2.1. With the sand, only microbial activity and micro-invertebrates (protozoa and metazoa) were measured.

5.2.3. Analytical methods

5.2.3.1. Water

Water analyses were performed following standard methods (APHA, 2012) unless specified otherwise. The samples were analyzed for pH and dissolved oxygen (DO) by electrometric methods. Chemical oxygen demand (COD) was analyzed using the open reflux titrimetric method, 5-d biological oxygen demand (BOD$_5$) by the oxygen electrode method, nitrite by the spectrophotometric method, ammonia by the dichloroisocyanurate method (NEN 6472, 1983), nitrates by the ion chromatograph (ICS 1000, DIONEX, USA), total suspended solids (TSS) by the gravimetric method and *E. coli* and fecal coliforms were determined by the plate count method using chromo cult agar (Merck, Germany). Total nitrogen (TN) was digested by the persulfate method and measured by the NO$_3^-$-N ultraviolet spectrophotometric screening method.

5.2.3.2. Sand and roots

The microbial activity of the biofilm attached to the sand and roots was assessed by the fluorescein diacetate (FDA) assay as described by Adam and Duncan (2001), but using 8 g of sample. To obtain the PNR and PDR of the sand, samples (~10 g) were put in 250 mL glass bottles containing 120 mL of the respective incubation solution following the method of Xu et al. (2013). For the PNR test, the incubation solution initial concentration was modified to ~300 mg NH$_4^+$-N L^{-1} (Appendix F) while the rest of ingredients were maintained as in Xu et al. (2013). The bottles were then placed on a rotary shaker at 175 rpm and 30°C in order to have favorable conditions for the nitrification process. For the PDR test, the incubation solution was identical to Xu et al. (2013) (~250 mg NO$_3^-$-N L^{-1}). The bottles were flushed with nitrogen gas (to create anaerobic conditions), capped and incubated at room temperature. Sampling for both tests was conducted at time 0 and 48 h and the nitrate concentration was analyzed. Results were calculated according to Xu et al. (2013). The PDR results were then multiplied by 85% as correction for some nitrate-ammonifiers present in the media converting NO_3^--N to NH_4^+-N (Zapater-Pereyra et al., 2015b - Chapter 8).

Micro-invertebrates (protozoa and metazoa) were determined by placing 20 g of sand in a plastic container, mixing it with ~10 mL of tap water and vortexing for ~10 s to dislodge the biofilm attached to the sand particles. From the suspension, 35 µL was transferred to a glass slide and covered with a cover slip. It was then observed under the reflected light fluorescence microscope with phase contrast and cell sense imaging software version 1.5 (Olympus, Germany). The different types of protozoa in the 35 µL were enumerated by grouping them into ciliates and flagellates, while for metazoa only rotifers were counted. Different taxonomic keys and guides by Finlay et al. (1988) were used for the in vivo identification of the cells to the genus level. The total population was expressed as population per mL (of tap water). Sand samples were analyzed within 12 h of sampling. For simplicity, sampling of sand in the VF CW was done when the compartment was drained. The HFF was permanently flooded, thus all the sampled sand included wastewater, but to maintain uniformity of the samples in the VF CW and HFF, this wastewater was discarded. Thus, for analysis, the sand of both compartments was mixed with tap water. Observations in the interstitial water showed that very few protozoa were present.

5.2.3.3. Batch experiment for protozoa and metazoa grazing on E. coli

The potential grazing of protozoa and metazoa on *E. coli* was investigated in a separate batch experiment. Sand from the Control VF CW (50 g) was added to two conical flasks and mixed with 150 mL of tap water. To assess the grazing effect of the micro-invertebrates on *E. coli,* the protozoa and metazoa population was inhibited in one flask with cycloheximide (0.03 g, reaching a concentration of 200 mg L^{-1} in each flask) following previous studies (Bomo et al., 2004; Tawfik et al., 2006; Chabaud et al., 2006). The flasks were kept open, placed on a rotary shaker at 80 rpm and at a temperature of 20°C. This experiment lasted 6 d and was conducted three times (n = 3). Sampling of *E. coli* from the flasks was done daily while protozoa and metazoa were enumerated every 3 d.

5.2.4. Data analysis

Data were analyzed using SigmaPlot 12.3 software. A one-way Analysis of Variance (ANOVA) followed by Tukey Post Hoc Test for all pairwise multiple comparison, at a 95% confidence interval (p=0.05), was used to compare the measured variables of the systems per sampling point (influent, along depth/length and effluents) and the microbial activity of the roots among the three systems. The Student T-test was used to compare the regular operation vs. the operation including external carbon source water to enhance denitrification. Parameters that did not meet assumptions were log_{10}-transformed; otherwise non-parametric ANOVA Kruskal-Wallis test was used.

5.3. RESULTS

5.3.1. **Removal of organic matter and solids**

The removal of BOD_5 was similar for the three investigated systems in both compartments (p>0.05; Fig. 5.2B). All BOD_5 was removed in the top part of the VF CWs; no further removal occurred (Fig. 5.2B). More than half of the COD concentration was removed in the top part of the VF CWs. The overall COD effluent concentrations were similar for all the three systems investigated (< 50 mg L^{-1}; p>0.05; Fig. 5.2C). The average DO concentration of all systems from all sampling points was ~2 mg L^{-1}, without showing significant differences among the systems (p>0.05). However, those in the Recirculating system were slightly higher compared to the other systems (Fig. 5.2A). The TSS greatly decreased in the top and middle part of all of the VF CWs to < 80 mg L^{-1} and in the HFF the concentrations decreased further to < 40 mg L^{-1} for all systems (p>0.05; Fig. 5.2D).

5.3.2. **Nitrogen**

The VF CWs removed almost all of the NH_4^+-N in all systems to the same extent (p>0.05; Fig. 5.2G). Once in the HFF, the Control and Recirculating systems showed lower NH_4^+-N concentrations than the Aerated system (Fig. 5.2G). NO_3^--N was formed in the VF CW of all the systems investigated. The HFF of the Control and Aerated system highly removed the formed NO_3^--N, but remained constant in the Recirculating system showing significant differences with the other systems (p<0.05; Fig. 5.2H).

TN concentrations slightly decreased in the VF CW compartment of all systems (Fig. 5.2I) showing no significant differences among the systems (p>0.05). In the HFF, the Control and Aerated systems showed better TN removal than the Recirculating system (p<0.05). The latter did not remove the TN concentrations further from the levels achieved by the VF CW (Fig. 5.2I). The nitrogen balance showed the same trend (Table 5.1): in the VF CW the order was Recirculating>Aerated>Control; for the HFF, it was Control>Aerated>Recirculating and in the whole Duplex-CW, it was Control=Aerated>Recirculating.

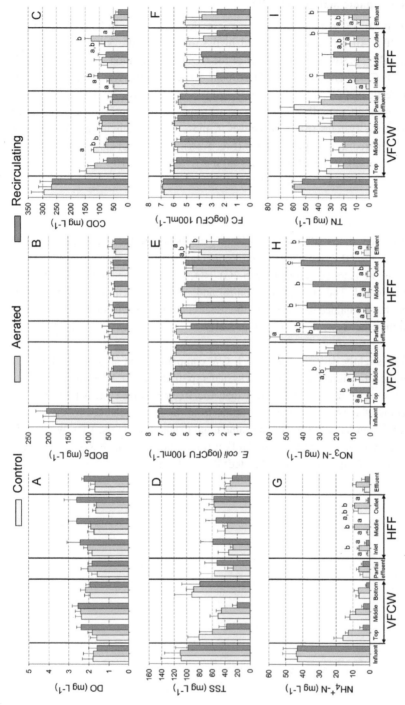

Figure 5.2 Water quality (Mean ± Std. error) along the depth of the vertical flow constructed wetland (VF CW) and the length of the horizontal flow constructed wetland (VF CW) and the length of the horizontal flow filter (HFF). Letters a,b,c indicate statistical significant differences between the systems, per sampling point. Only significant differences were displayed. FC = fecal coliforms.

Table 5.1 Total nitrogen balance (of one feeding day, 39 L of wastewater) in the vertical flow constructed wetland (VF CW) and horizontal flow filter (HFF) of all tested Duplex-CWs.

Duplex-CW	Influent mg	Partial effluent mg	Effluent mg	VF CW removal mg / g m^{-2} d^{-1}	HFF removal mg / g m^{-2} d^{-1}	Duplex-CW removal mg / g m^{-2} d^{-1}
Regular operation						
Control	2067	2301	273	-234 / -1.0	2028 / 2.4	1794 / 1.7
Aerated	2301	1482	507	819 / 3.4	975 / 1.2	1794 / 1.7
Recirculating	2067	1170	1248	897 / 3.7	-78 / -0.1	819 / 0.8
Operation with the addition of filtered influent to enhance denitrification						
Control	-	2510*	429	-	2081 / 2.5	-
Recirculating	-	1999*	1521	-	478 / 0.6	-

*Considers the nitrogen from the filtered influent added.
Negative values indicates that the outlet concentration was higher than inlet concentration.

The PNR of biomass present was higher in the VF CWs than that of the HFFs (Table 5.2). Among all, the Recirculating system had higher PNRs than the other systems. In general, the PDR was higher in the HFF than in the VF CW. The Aerated HFF showed the highest PDR (42 mg N kg^{-1}$_{sand}$ h^{-1}; Table 5.2). Comparison of the average values of PNR and PDR with the actual nitrification and denitrification rates (ANR and ADR) showed that the ANR were in the same range of the PNR in all VF CWs, but the ADR were much lower than the PDR (Table 5.2). For the HFF, both the ANR and ADR were orders of magnitude lower than the predicted PNR and PDR, respectively (Table 5.2).

5.3.3. Carbon source as electron donor for denitrification

The water quality of the filtered influent is given in Table 5.3. The performance of the Control and Recirculating systems was similar with and without filtered influent for COD, TSS, NH_4^+-N, NO_3^--N, TN, *E. coli*, fecal coliforms, microbial activity, protozoa and metazoa (Table 5.3). The application of filtered influent decreased the BOD$_5$ concentrations in both systems, however only the Recirculating system showed significant differences at the effluent (Table 5.3). The DO concentration in the Control system was lower without (1-3 mg L^{-1}) than with filtered influent (2-5 mg L^{-1}), however the effluent concentration did not show significant differences (p>0.05, Table 5.3). The opposite effect occurred with the Recirculating system (Table 5.3). The nitrogen removal rate in the Recirculating system highly increased (from -0.1 to 0.6 g m^2 d^{-1}), while no changes were observed in the Control system (Table 5.1).

Table 5.2 Actual *vs.* potential nitrification and denitrification rates (ANR *vs.* PNR and ADR *vs.* PDR) in the vertical flow constructed wetland (VF CW) and horizontal flow filter (HFF) of all tested Duplex-CWs and of other studies. Units in mg N kg$^{-1}_{sand}$ h^{-1}.

Constructed wetland	Reference	VF CW				HFF			
		ANR*	PNR (n = 6)	ADR** (n = 6)	PDR (n = 6)	ANR*	PNR (n = 6)	ADR** (n = 6)	PDR (n = 6)
Control Duplex-CW	This study	0.27	0.23	-0.05	29	0.005	0.13	0.24	33
Aerated Duplex-CW	This study	0.26	0.24	0.15	23	-0.008	0.19	0.11	42
Recirculating Duplex-CW	This study	0.27	0.30	0.16	24	0.007	0.16	-0.01	36
Control Duplex-CW***	Zapater-Pereyra et al. 2015a	0.57	~0.04	0.30	-	0.013	-	0.19	-
VF CW with *P. australis*	Xu et al. 2013	-	~-0.04	-	~11	-	-	-	-

* Calculated for NH$_4^+$-N.
** Calculated for total nitrogen.
*** Identical operation as the control Duplex-CW in this study but with higher wastewater strength (Fill and drain configuration from Chapter 6, Fig. 6.4). PNR and PDR are the average of the three measurements along depth of the VF CW and length of the HFF.

Table 5.3 Concentrations of the various water quality parameters of the Control (C) and Recirculating (R) horizontal flow filters (HFF) with and without the use of filtered influent as electron donor to enhance denitrification.

Parameter	Duplex-CW	Filtered influent (Mean ± Std. dev.)	Without filtered influent		With filtered influent		P-value[b]
			HFF range[a]	HFF effluent (Mean ± Std. dev.)	HFF range[a]	HFF effluent (Mean ± Std. dev.)	
DO (mg L^{-1})	C	3 ± 1	1 - 3	2 ± 1	2 - 5	3 ± 2	0.174
	R		1 - 5	2 ± 0	2 - 5	4 ± 1	**0.029**
BOD$_5$ (mg L^{-1})	C	142 ± 28	21 - 64	31 ± 7	9 - 32	18 ± 8	0.070
	R		24 - 59	33 ± 8	11 - 33	17 ± 6	**0.037**
COD (mg L^{-1})	C	181 ± 24	24 - 155	46 ± 18	23 - 69	38 ± 6	0.614
	R		16 - 151	33 ± 19	35 - 90	48 ± 11	0.287
TSS (mg L^{-1})	C	139 ± 34	5 - 121	38 ± 20	12 - 83	26 ± 18	0.424
	R		4 - 112	27 ± 39	20 - 88	40 ± 10	0.215
NH$_4^+$-N (mg L^{-1})	C	46 ± 1	0 - 16	3 ± 2	1 - 5	3 ± 1	0.723
	R		0 - 11	3 ± 5	0 - 1	0 ± 0	0.071
NO$_3^-$-N (mg L^{-1})	C	0.06 ± 0.02	1 - 7	4 ± 3	0 - 14	6 ± 4	0.414
	R		23 - 53	38 ± 8	33 - 45	38 ± 4	0.951
Total nitrogen (mg L^{-1})	C	60 ± 2	1 - 37	7 ± 4	3 - 18	11 ± 6	0.306
	R		2 - 50	32 ± 19	37 - 47	39 ± 3	0.540
E. coli (CFU 100ml^{-1})	C	$3\times10^6 \pm 1\times10^6$	$0 - 5\times10^5$	$7\times10^4 \pm 9\times10^4$	$4\times10^3 - 2\times10^5$	$7\times10^4 - 6\times10^4$	0.786
	R		$0 - 4\times10^5$	$6\times10^3 \pm 5\times10^3$	$4\times10^3 - 6\times10^5$	$2\times10^4 - 2\times10^4$	0.250
Faecal coliforms (CFU 100ml^{-1})	C	$4\times10^6 \pm 1\times10^6$	$5\times10^4 - 3\times10^5$	$2\times10^5 \pm 8\times10^4$	$4\times10^3 - 2\times10^5$	$4\times10^4 - 5\times10^4$	0.067
	R		$0 - 4\times10^5$	$8\times10^4 \pm 1\times10^5$	$3\times10^3 - 1\times10^5$	$2\times10^4 - 2\times10^4$	1.000
Microbial activity (μg$_{fluorescein}$ g$^{-1}_{sand}$ h^{-1})	C	-	6 - 21	-	8 - 40	-	> 0.05
	R	-	3 - 19	-	3 - 28	-	> 0.05
Protozoa and metazoa (Polulation mL^{-1})	C	-	429 - 943	-	257 - 714	-	> 0.05
	R	-	114 - 686	-	257 - 629	-	> 0.05

[a] Include all results obtain within the HFF profile (inlet, middle and outlet) and the HFF effluent.
[b] P-values correspond to the comparison of the HFF effluents with and without the use of filtered influent, except for microbial activity and protozoa and metazoa which correspond to p-values obtained at all sampling points within the bed media. Bold letters indicate significant differences (p<0.05).

5.3.4. **Pathogens, protozoa and metazoa**

E. coli and fecal coliforms were removed in all systems from 2 to 5 logs. All VF CWs removed up to 2 logs, while the HFF removed further up to 3 logs (Fig. 5.2E-F). The Recirculating system showed the highest pathogen removal efficiency (Fig. 5.2E-F). However, in almost all the cases there were no significant differences at each sampling point between the systems ($p>0.05$).

In all Duplex-CWs, ciliates were the most abundant, followed by the flagellates and lastly the rotifers (Fig. 5.3B). For the Control system, the numbers of the protozoa and metazoa population decreased from top to bottom of the VF CW and increased along the length of the HFF. The Aerated system showed no trend in the VF CW, while in the HFF it increased from the inlet to the outlet. The Recirculating system showed, among all systems, the largest population at the top of the VF CW (~700 population mL^{-1}), which gradually decreased towards the bottom and then remained constant in the HFF. Despite visible changes, there were no differences among the systems at each sampling point ($p>0.05$; Fig. 5.3B).

Grazing of protozoa and metazoa on *E. coli* (Fig. 5.4) showed that in the presence of protozoa, metazoa and *E. coli* (solid line, Fig. 5.4), the bacterial numbers decreased gradually, while the size of the protozoa and metazoa population increased (within 3 d) and then decreased till the end of the experiment. When cycloheximide was added (dashed line, Fig. 5.4), the protozoa and metazoa population was slowly reduced while it was visible that the *E. coli* population maintained along the experiment (Fig. 5.4).

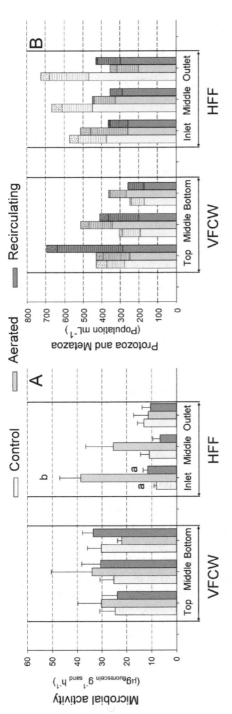

Figure 5.3 Microbial activity (Mean ± Std. error) (A) and protozoa and metazoa counts (B) along the depth and length of the vertical flow constructed wetland (VF CW) and horizontal flow filter (HFF), respectively. In Graph B, each pattern indicates the type of organisms: none, ciliates; dashed, flagellates and dotted, rotifers. Letters a,b indicate statistical significant differences between the systems, per sampling point. Only significant differences were displayed. *Note: to express the microbial activity per volume (mL_sand), values displayed in the figure should be divided by 0.65 mL g^{-1}.*

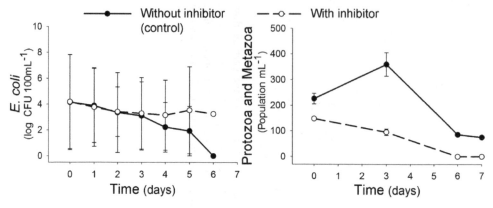

Figure 5.4 Protozoa and metazoa grazing on *E. coli* batch experiment (Mean ± Std. error).

5.3.5. Microbial activity

The activity of the microbial community attached to the sand was, in general, higher in the VF CW than in the HFF for all systems investigated (Fig. 5.3A). Only at the inlet and middle of the HFF of the Aerated system, the microbial activity was similar to that in the VF CW, it then decreased to levels similar to the other systems (Fig. 5.3A). For almost all the cases, no significant differences occurred among the systems ($p > 0.05$; Fig. 5.3A). The microbial activity on the roots was much higher than that on the sand (Mean ± Std. error): 592 ± 174 $\mu g_{fluorescein}$ g^{-1} $_{root}$ h^{-1}, 2977 ± 1244 $\mu g_{fluorescein}$ g^{-1} $_{root}$ h^{-1} and 3240 ± 1152 $\mu g_{fluorescein}$ g^{-1} $_{root}$ h^{-1} for the Control, Aerated and Recirculating Duplex-CWs, respectively. In volume terms the mean activity on the roots was: 49 $\mu g_{fluorescein}$ mL^{-1} $_{root}$ h^{-1}, 246 $\mu g_{fluorescein}$ mL^{-1} $_{root}$ h^{-1} and 267 $\mu g_{fluorescein}$ mL^{-1} $_{root}$ h^{-1}. No significant differences between the systems investigated were observed ($p > 0.05$).

5.4. DISCUSSION

5.4.1. Role of recirculation and aeration in the Duplex-CW design

Recirculation was used in this study to enhance the treatment performance by extending the contact time between the wastewater and the media (Sun et al., 2003; Foladori et al., 2014) and to provide additional electron donor for denitrification from remaining interstitial water and roots exudates (Zhai et al., 2013). Recirculation can enhance bed oxygenation as well, which is also the main purpose of artificial aeration (Foladori et al., 2013, 2014). In spite of those theoretical benefits, the intensified systems performed similar to the Control Duplex-CW (Figs. 5.2 and 5.3; Tables 5.1 and 5.2). Recirculation slightly enhanced the water DO in the HFF, nevertheless it was not significantly different to the other systems ($p > 0.05$, Fig. 5.2A).

Literature has reported the benefits on the CW performance brought by aeration and recirculation, mainly for organic matter and NH_4^+-N (Table 5.4). Such studies investigated

HF CWs (known for their limited oxygen availability) and/or highly loaded CWs (6-120 g COD m^{-2} d^{-1} and 63-270 L m^{-2} d^{-1}, Table 5.4). In this study, the use of two compartments to treat low loads (13 g COD m^{-2} d^{-1} and 46 L m^{-2} d^{-1}, Table 5.4) were sufficient to provide an appropriate treatment without triggering the effects of aeration or recirculation. Fan et al. (2013a) mentioned that conventional wetlands were always efficient in organic matter removal at low loading rate. They found that at 8 g COD m^{-2} d^{-1}, the removal of COD and TN did not differ significantly between aerated and non aerated VF CW (aerated: 91%, 26% and non aerated: 83%, 39%, respectively). However, when the OLR was increased to 59 g COD m^{-2} d^{-1}, the benefit of aeration became more obvious (over 30% difference) (Table 5.4). Also Butterworth et al. (2013) did not find any benefit of aeration in the removal of carbonaceous BOD_5 in a HF CW due to the low loads applied (19 g COD m^{-2} d^{-1}) despite an applied HLR 6-fold higher than in this study (270 L m^{-2} d^{-1}, Table 5.4).

Despite the low loads applied, BOD_5 and COD concentrations remained above 30 mg L^{-1} (Fig. 5.2B-C). It has been reported that unsaturated VF CWs usually obtain concentrations of BOD_5 below 20 mg L^{-1} (Boog et al., 2014) and that background concentrations for BOD_5 are ~ 2 mg L^{-1} (Kadlec and Wallace, 2009). This might be due to the short HRT of the VF CWs (explained in Section 5.4.2) and the low efficiency of the air bubbles to flow through the sand medium. Artificially aerated systems are characterized by a high effluent DO (3-11 mg L^{-1}), but in this study it was only ~2 mg L^{-1} (Fig. 5.2A, Table 5.5). This can be related to the higher air flow applied but, for example, Fan et al. (2013a,b) applied one third of the air flow used in this study and still their effluent was saturated with oxygen (Table 5.5). Therefore, at the end of this study, in an attempt to verify the air bubbles distribution through the sand profile, all the VF CWs were flooded with tap water and the air blower was switched on (artificial aeration installations were included in all the systems). It was observed that the bubbles were not properly distributed all over the area, but only in random points near the edges of the system, mainly close to the position where the aeration pipe enters vertically to the system (Appendix E). This situation likely decreased the chances of bed oxygenation that could have increased the organic matter removal in the Aerated system.

Kadlec and Wallace (2009) recommend the use of aeration only when its costs (operation and maintenance) compensate a reduction in capital costs (e.g. requiring a smaller CW footprint resulting in a reduced cost per m^2 of land and into a smaller construction cost). The first order kinetic equation (assuming no background concentration) revealed that the ratio of the Control area to the Aerated or Recirculating system area was 0.3-1.3, which means that the area achieved by the intensified systems was similar or slightly larger than that of the Control (Appendix C). The use of intensification (i.e. aeration and recirculation) is thus not recommended in the Duplex-CWs for the operational conditions applied in this study.

Table 5.4 Comparison of the organic (OLR) and hydraulic loading rates (HLR) between studies using intensified vertical (VF) and horizontal flow (HF) constructed wetlands (CW) and the benefits introduced by the use of intensification as compared to a Control system.

CW type	Intensif.	OLR (g COD m⁻² d⁻¹)	HLR (L m⁻² d⁻¹)	Benefit (C: Control, A: Aerated, R: Recirculating)			Reference
				Organic matter	Nitrogen	Others	
Aeration							
VF	Continuous[a]	13	46	None	None	-	This study
HF	Continuous	11-20	68	COD (A: 89%, C: 69%)	NH_4^+-N (A: 99%, C: 9%)	Microbial activity Protozoa abundance Solids accumulation Area reduction	Zapater-Pereyra et al. 2014
VF[b]	Continuous/ Intermittent	27[c]	95	cBOD5 (A: >99%)	TN (A: 60% / 78%)	-	Boog et al. 2014
HF	Continuous	19	270	None for cBOD5	NH_4^+-N (A: 99%, C: 13%)	Solids accumulation	Butterworth et al. 2013
VF	Intermittent	59	70	COD (A: 97%, C: 66%)	NH_4^+-N (A: 96%, C: 23%) TN (A: 94%, C: 30%)	-	Fan et al. 2013a
VF	Intermittent	64	158	COD (A: 88%, C: 80%)[d]	TN (A: 49%, C: 29%)[d] None for NH_4^+-N	Area reduction	Foladori et al. 2013
VF	Continuous/ Intermittent	12-120	190-760	COD (A: 80-72% / 78-64%, C: 57-48%)	NH_4^+-N (A: 87-65% / 78-54%, C: 61-40%) None for TN	-	Dong et al. 2012
Recirculation							
VF	Continuous[a]	13	46	None	None	-	This study
VF	Intermittent	83	168	COD (R: 85%, C: 80%)[d]	TN (R: 44%, C: 29%)[d] None for NH_4^+-N	Area reduction	Foladori et al. 2013
HF-VF[e]	Continuous	6-12	126	Not measured	TN (R: 66%, C: 29%)	-	Ayaz et al. 2012
VF (3 stages)	Continuous	~74	63	BOD5 (R: 97%, C: 72%) COD (R: 78%, C: 51%)	NH_4^+-N (R: 70%, C: 19%)	TSS (R: 91%, C: 44%) None for PO_4^{3-}-P	Sun et al. 2003
Aeration and Recirculation							
VF	Intermittent	74	179	COD (A+R: 92%, C: 80%)[d]	TN (A+R: 75%, C: 29%)[d] None for NH_4^+-N	Area reduction	Foladori et al. 2013

[a] During the 1 d hydraulic retention time.
[b] No Control system was used in this study.
[c] COD was not reported, the organic loading rate displayed is based on carbonaceous BOD5 (cBOD5).
[d] Applied loads to the Control system (32 g COD m⁻² d⁻¹ and 69 L m⁻² d⁻¹) were lower than those in the Aerated system.
[e] Reported values stand for the tested "100% recirculation".

Table 5.5 Air flow provided by the bottom aeration and effluent dissolved oxygen (DO) concentrations in different vertical (VF) and horizontal flow (HF) constructed wetlands (CW).

CW type	Aeration type	Filter media (mm)	Air flow ($m^3 h^{-1}$)	Effluent DO ($mg L^{-1}$)	Reference
VF	Continuous[a]	70 cm Sand (1-2)	0.3	~2	This study
VF (saturated)	Continuous / Intermittent	80 cm Gravel (8-16)	2.2	8.1 / 6.3	Boog et al. 2014
HF	Continuous	38 cm Gravel (8-16)	0.45-1.4	6.8-7.7	Zapater-Pereyra et al. 2014
HF	Continuous	60 cm Gravel (6-12)	150	8-11 (within bed)	Butterworth et al. 2013
VF	Intermittent	15 cm Gravel (<2) 15 cm Gravel (10-20) 25 cm Gravel (10-30)	~0.09	8.18[b]	Fan et al. 2013a,b
VF	Continuous / Intermittent[c]	50 cm Gravel (7-15)	NA	4.4 / 3	Dong et al. 2012

NA, Not available.
[a] During the 1 d hydraulic retention time.
[b] Reported values are from the tested system with COD/N ratio of 2.5 (the closer ratio to that one of the Duplex-CW (1.45)).
[c] Aeration was not at the bottom of the system but at the middle. So only the upper half of the CW was aerated. DO results correspond to the aerated half.

5.4.2. Organic matter and nitrogen removal

Typically in a VF CW, immediately after an intermittent feeding, the water oxygen levels increase ($\sim$ 7.5 mg L^{-1}) and within few hours fast oxygen depletion occurs (Zapater-Pereyra et al., 2015a - Fig. 6.6, Chapter 6). DO depletion (to < 0.61 mg L^{-1}) has been reported within 4 h (Fan et al., 2013a,b). Therefore, the VF CW of the Duplex-CWs were likely saturated with oxygen and the majority of it was consumed during organic matter biodegradation and nitrification. This was confirmed by the majority of COD, BOD_5, TSS and NH_4^+-N consumed in that compartment (Fig. 5.2).

Since the intensification did not contribute significantly to the treatment due to the low loads applied, it is reasonable to assume that the similar organic matter, solids and NH_4^+-N removal in all systems occurred with the intrinsic DO (Fig. 5.2). The effluent DO ($\sim$2 mg L^{-1}, Fig. 5.2A) represented only $\sim$0.09 g O_2 m^{-2} d^{-1} (calculated as: 2 mg L^{-1} $\times$ 46 L m^2 d^{-1}), but the oxygen consumption rate was 18 g O_2 m^{-2} d^{-1} for the VF CWs and 0-1 g O_2 m^{-2} d^{-1} for the HFFs, calculated from the maximum oxygen consumption rate equation from Nivala et al. (2013) using BOD_5 and NH_4^+-N values. It can therefore be extrapolated that the measured DO did not reflect the VF CW aerobic environment. Effluent DO is not a good indicator of aerobic/anaerobic conditions within the CW (Vymazal and Kröpfelová, 2008). For the HFF, the similarity between both values suggested almost no depletion of oxygen in that compartment and confirmed its main anoxic role.

The three Duplex-CWs effluent met concentration guidelines for COD, but not for BOD_5 (125 and 25 mg L^{-1}, respectively; EEC, 1991). However, the BOD_5 load removal (79-84%) achieved the requirements specified in the same guidelines (70-90%, EEC, 1991). Nevertheless, BOD_5 and COD concentrations remained > 30 mg L^{-1} despite the fact that VF CWs are able to provide lower BOD_5 concentrations (Boog et al., 2014) and that the majority of the available carbon was biodegradable (COD:BOD_5 ratio of 1.45 and 1.23, respectively), as expected from municipal wastewater (Vymazal and Kröpfelová, 2009). The total Duplex-CW HRT was 4-5 d, however only 1 d was associated with the unsaturated VF CWs. Knowing that aerobic organic matter degradation is usually more important (Kadlec and Wallace, 2009; Dong et al., 2012) and that BOD_5 removal is critical below a HRT of 1 d (Weerakoon et al., 2013), it can only be assumed that the short HRT (1 d) limited the further BOD_5 (and COD) removal. Prochaska et al. (2007) found a COD removal of > 92% in a sand based VF CW fed with HLR of 23-73 L m^{-2} d^{-1} and with HRT of 1.4-2.9 d, while in this study the COD removal efficiency was 75-81% in the VF CW with 1 d HRT, and only increased up to 83-88% with the extra 3-4 d provided by the HFF. Ghosh and Gopal (2010) demonstrated that a HRT of 4 d in a saturated VF CW enhanced organic matter and nitrogen removal as compared to a HRT of 1 d. Further experiments should be conducted to define an optimal HRT for the VF CW, without compromising the total Duplex-CW HRT. Considering that the DO is already highly depleted after 1 d, intensification might become necessary when extending the HRT.

Nitrogen removal (via denitrification) took place in the VF CWs (after oxygen depletion) of all systems and in the HFFs of only the Control and Aerated systems (Fig. 5.2I and Table 5.1). The minor difference in the DO concentration between the HFF of the Recirculating and the other systems appeared to impede denitrification (Fig. 5.2H-I), but it was not enough to further decrease the remaining organic matter concentration (> 30 mg L^{-1}, Fig. 5.2B-C). According to Kadlec and Wallace (2009), denitrification can occur in low oxygen environments but not above 0.3-1.5 mg L^{-1}. Hiscock et al. (1991) mentioned that denitrification cannot occur at a DO > 0.2 mg L^{-1}. In this study, denitrification occurred partially in all compartments (Fig. 5.2I), except in the Recirculating HFF that showed slightly higher DO concentration (1-5 mg L^{-1}; Table 5.3 and Fig. 5.2A). Therefore, TN regulations (15 mg L^{-1}) (EEC, 1991) were only achieved by the Control and Aerated systems.

5.4.3. Filtered influent as carbon source to enhance denitrification in the HFF

The most important factors to achieve denitrification of available NO_3^--N are the presence of organic carbon as electron donor and an anoxic environment (Fan et al., 2013a). Despite the partial effluent in the Recirculating system already contained biodegradable carbon (BOD_5 = 49 mg L^{-1} and $COD:BOD_5$ = 1.1, Fig. 5.2B-C), in this experiment extra carbon was added to test its role in denitrification. In general, the addition of filtered influent did not introduce changes in the performance of the Recirculating Duplex-CW, except slightly for DO and BOD_5 concentrations (Table 5.3). Manipulation of the filtered influent (i.e. filtering, addition to the bucket and mixing with the partial effluent) oxygenated further the Recirculating and Control HFFs and probably that was the reason for the slightly higher BOD_5 removal (Table 5.3).

The non significant effluent TN concentrations with and without the use of filtered influent in both systems (Recirculating and Control) (p>0.05, Table 5.3), indicated that the low capacity observed in the Recirculating Duplex-CW to denitrify nitrogen was not related to a carbon deficiency but to the oxic conditions that only allowed a partial denitrification (Fig. 5.2A and Table 5.3). Thus, the load increment of denitrified nitrogen from -0.1 to 0.6 g m^{-2} d^{-1} (Table 5.1) could be related to the larger amount of nitrogen introduced to the HFF with the filtered influent (1170 vs. 1999 mg N, Table 5.1), since the higher the influent concentration the higher the TN removal rates. An extra carbon source was bypassed to the HFF (Hybrid-B, Chapter 4) to enhance denitrification in the study of Zapater-Pereyra et al. (2014) (Chapter 4), however the reduced conditions of the tested HF CWs (0.9 mg O_2 L^{-1}, Table 4.1) did not allow the bypass water to be nitrified, therefore denitrification was offset by the lack of nitrification.

5.4.4. Role of the Duplex-CW compartments in nitrogen removal

The VF CWs showed higher PNR than the HFFs (Table 5.2) due to the resting periods that allowed re-aeration of the bed (Zapater-Pereyra et al., 2015a - Chapter 6), while the PDR was higher in the HFFs than in the VF CWs (Table 5.2) due to the saturation conditions. This reconfirmed the main function of the VF CWs, as explained in Section 5.4.2 and shows that the microbial community in the HFF has the potential to denitrify to levels higher than those in the VF CWs if anoxic conditions occur (Table 5.2).

The PNR and PDR batch experiments intend to provide (in a flask) the ideal conditions (e.g. temperature, dissolved oxygen and C:N ratio) for nitrification and denitrification, respectively. Both were measured also to predict the maximum potential capacity of each compartment to remove nitrogen if conditions are optimized. In the HFFs the ADR was 100-fold lower than the PDR, showing that the HFF is not working to its maximum capacity (overdesigned). Denitrification is the main NO_3^--N (TN) removal mechanism in CWs (Saeed and Sun, 2012) that usually is limited by the high DO concentrations (as in this study) or an electron donor. In the PDR methodology those limitations did not exist, therefore, for all systems tested, the PDRs showed that the HFF could enhance the TN removal by improving the conditions for denitrification.

The small differences between the ANR and PNR values in the VF CWs (Table 5.2) indicated that all VF CWs were used to their maximum capacity. This implies that the VF CW would not be able to treat higher loads while the HFF will, if optimal conditions (as in the flask) would exist in the CW. This observation was incorrect for the PNR, since the ANR of a VF CW from an identical Control Duplex-CW (fill and drain configuration studied in Chapter 6) treating domestic wastewater with higher strength (Table 5.2; Zapater-Pereyra et al., 2015a) was higher than the predicted PNR (0.57 vs. 0.23-0.30 mg N kg^{-1}_{sand} h^{-1}; Table 5.2), showing that the Duplex-CW was capable of reaching higher rates than what the potential values could predict. This highlights the limitations of predicting values using PNR test probably because the batch experiments do not consider other nitrogen removal processes that occur in a CW such as plant uptake, adsorption and biomass assimilation. The latter two processes can occur in the batch experiment but the quantities of the material used in the batch experiments are not proportional to reality.

The PNR and PDR from another VF CW with a HRT of 2 d (Xu et al., 2013) were smaller than those obtained in the studied VF CWs (Table 5.2). This could be due to the slight differences between the used PNR and PDR methodologies. Preliminary studies done for the PNR methodology (Appendix F) suggested that results varied depending on the incubation solution NH_4^+-N concentration. The higher the NH_4^+-N concentration, the lower the PNR result (Appendix F). On the other hand, the PDR methodology used in this study guaranteed anoxic conditions by flushing each flask with nitrogen gas prior initiating the experiment. Xu et al. (2013) worked with a higher PNR concentration than in this study and without the use of nitrogen gas. Furthermore, the g of sample per mL of solution also varied greatly (1:1 in Xu et al. (2013) and 1:12 in this study).

Different PNR and PDR methodologies or variations within the same methodology will lead to deviations in the results and are thus, impossible to compare among studies. Furthermore, due to the characteristics of the tests, they can be a tool to predict potential values that can be achieved when the CW microbial community is subjected to the process-specific optimal conditions (e.g. nitrification or denitrification), but any forecasted value is strictly linked to the methodology applied. Therefore, extrapolations using the maximum potential capacity should be avoided. Further investigation is required for those methodologies in order to understand their capability to predict maximum nitrification and denitrification rates if optimal conditions would exit in the CW. This was out of this thesis scope. The methods are still considered be valuable for comparison and to show trends (as in Chapter 8) only among results obtained with the same methodology.

5.4.5. Bacteria and protozoa

The *E. coli* and fecal coliforms removal in the Duplex-CW (2-5 logs) was similar to removal efficiencies found in other CWs (Kadlec and Wallace, 2009). García et al. (2013) found a removal of 3-4 logs for both coliforms and *E. coli* using a hybrid system (VF-HF sequence). Discharging the Duplex-CW effluent is possible since pathogens are not included in the European discharge guidelines (EEC, 1991), but for agricultural reuse purposes (USEPA, 2012) the pathogen removal by the Duplex-CW (and by CWs in general) is yet not enough. Therefore, depending on the final reuse purpose disinfection might be needed.

The slightly higher bacterial removal in the HFF as compared to the VF CW of all systems (Fig. 5.2E-F) was the result of the HRT. Bacteria in subsurface flow CWs are removed via predation, settling and filtration by plants and filter media (Kadlec and Wallace, 2009). All of those processes are time dependent, thus the 3-4 d HRT in the HFF enhanced the bacterial removal as compared to the 1 d HRT in the VF CWs. The time dependency of predation was confirmed by the grazing experiment (Fig. 5.4) that showed that the longer the exposure, the larger the bacterial removal up to a certain bacterial concentration (solid line in Fig. 5.4), in agreement with the literature (e.g. Decamp and Warren, 1998; Kadlec and Wallace, 2009; García et al., 2013).

Usually protozoa populations augment at increasing DO (Fig. 4.5, Zapater-Pereyra et al., 2014). In this study, no significant differences in protozoa population existed at each sampling point between the systems (p>0.05; Fig. 5.3B) despite the hypothesis that the VF CWs had more oxygen that the HFF (Section 5.4.2). Furthermore, the slightly higher DO in the Recirculating HFF was not reflected in the protozoan counts (Fig. 5.3B). Nevertheless, metazoa (rotifers) and protozoa, mainly ciliates, were present in the Duplex-CW compartments and were actively grazing on bacteria (Fig. 5.4) but *E. coli* removal depends on different mechanisms and grazing alone will never achieve a complete bacterial removal since bacteria growth rate exceeds predation rates (Matz and Kjelleberg, 2005). For example, Decamp and Warren (1998) found that the maximum grazing rate by one cell of the ciliate *Paramecium* was 100 *E. coli* h^{-1}. A very rough calculation, assuming that the average protozoa and metazoa population in the Duplex-CWs will graze at that rate, showed that all *E. coli* could have been removed within 1 h in this study, which did not occur (Fig. 5.2E). This is because, despite ciliates are a majority in the Duplex-CWs (Fig. 5.3B) and free-swimming ciliates (like *Paramecium*) are the most abundant among the ciliates in the CWs (Fig. 4.5, Zapater-Pereyra et al., 2014), not all other protozoa and metazoa existing in CWs preserve the same (active) population size, graze on (suspended) bacteria and at the same rate, or graze constantly independent of factors such as temperature, prey concentration (Fig. 5.4) or redox potential (Decamp and Warren, 1998).

5.4.6. Microbial activity

Aerobic conditions enhance microbial activity (Fig. 4.4, Chazarenc et al., 2009; Zapater-Pereyra et al., 2014). This explains the overall higher activity in the oxic VF CWs (Section 5.4.2) as compared to the HFFs (Fig. 5.3A). Since intensification did not create DO significant differences between the systems, as well no differences were encountered in the activity ($p > 0.05$; Fig. 5.3A). The higher activity at the inlet of the Aerated HFF could have been caused by some air bubbles travelling towards the HFF, since the large voids (gravel drainage layer and pipes) along the bubble path could facilitate air transport. Surprisingly, this was not visible in other parameters, i.e. DO, organic matter, NH_4^+-N and protozoa (Figs. 5.2 and 5.3) at the same sampling point. The microbial activity values encountered in the sand of the Duplex-CW (Fig. 5.3A) where in the lower half of those measured in the substratum of a shallow (9 cm depth) HF CW called the constructed wetroof (16-76 $\mu g_{fluorescein}$ $g^{-1}_{substratum}$ h^{-1}; Fig. 8.3). The highly aerobic conditions in the constructed wetroof (effluent DO ~ 5-6 mg L^{-1}; Table 8.2) enhanced the activity.

The microbial activity in the roots (49-267 $\mu g_{fluorescein}$ mL^{-1}_{root} h^{-1}) was by far higher than that of the sand (10-59 $\mu g_{fluorescein}$ mL^{-1}_{sand} h^{-1}), due to their oxygen release and exudates excreted, that contribute to the development of active biofilm (Weerakoon et al., 2013; Zhai et al., 2013). Nevertheless, in absolute terms, the microbial community in the filter media governs the treatment, since there are always more filter media than roots in a CW. Calculations indicate an absolute activity of 1.7-9.9 g h^{-1} for sand (calculated as: 10-59 $\mu g_{fluorescein}$ mL^{-1}_{sand} h^{-1} × 0.24 m^2 × 0.7 m) and 0.06-0.34 g h^{-1} for the roots (calculated as: 49-267 $\mu g_{fluorescein}$ mL^{-1}_{root} h^{-1} × 106 g / 0.083 g mL^{-1}). The total root volume of the VF CWs (106 g) was taken from Zapater-Pereyra et al. (2015a) for a similar CW but after 435 d of operation (Chapter 6). Probably these values would be even lower in this study as the plants (roots) grew for about 150 d.

5.5. CONCLUSIONS

▪ All studied Duplex-CW systems met European COD and BOD_5 discharge guidelines. Nevertheless, concentrations remained above 30 mg L^{-1} due to the short HRT in the VF CW. Only the Control and the Aerated systems met the TN requirements. Pathogen removal was significant (up to 5 logs), but not enough to meet reuse guidelines for agriculture.
▪ The applied loads did not trigger the benefits of aeration and recirculation.
▪ The VF CW treated mainly organic matter, solids and NH_4^+-N and to a lesser extent pathogens and TN. The HFF contributed to a further pathogen removal and to TN removal (except in the Recirculating system).
▪ Application of filtered influent to the Recirculating HFF as a carbon source to enhance denitrification was not able to enhance the TN removal because the DO concentrations were too high.

CHAPTER 6.

EVALUATION OF THE PERFORMANCE AND SPACE REQUIREMENT BY THREE DIFFERENT HYBRID CONSTRUCTED WETLANDS IN A STACK ARRANGEMENT

Lack of space for a wastewater treatment plant is a common problem in many areas, especially in dense cities. Constructed wetlands (CWs) are efficient natural systems; however they require large areas for an appropriate wastewater treatment. The aim of this study is the development of a compact CW design for the treatment of domestic wastewater, the Duplex-CW: a hybrid system with a stacked design (vertical flow CW (VF CW) on top of a horizontal flow filter (HFF)). The performance of three different configurations of Duplex-CW, called Fill and drain, Stagnant batch and Free drain, was compared. The VF CWs operated differently with the intention of creating different oxygen conditions, whereas the HFFs were operated identically. The Duplex-CWs were subjected to three different wastewater strengths, corresponding to designs of 7.9, 3.4 and 2.6 m^2 PE^{-1}. The highest strength was treated with and without artificial aeration of the VF CW of each configuration. The contribution to the total removal of each compartment (VF CW and HFF), the effects of the use of artificial aeration, the solids accumulation, above- and below-ground biomass and the footprint requirements of the three configurations tested were determined. The Fill and Drain configuration performed better than the other two, the VF CW compartment being more active in the treatment than the HFF. It achieved an area of 2.6-3.4 m^2 PE^{-1} and it needed 2-3 times lower area than what a single VF CW would have needed to reach similar total nitrogen effluent concentrations. The Duplex-CW did not contribute to the footprint reduction for other parameters (e.g. COD, TSS and total phosphorus).

This chapter is based on:
Zapater-Pereyra M., Ilyas, H., Lavrnić S., Bruggen van J.J.A., Lens P.N.L. (2015), "Evaluation of the performance and the space requirement by three different hybrid constructed wetlands in a stack arrangement", Ecological Engineering, 82, 290-300.

6.1. INTRODUCTION

Constructed wetlands (CWs) are engineered to mimic natural wetlands and efficiently remove a wide range of pollutants (mainly organic matter) from wastewater. In certain situations, their usage is limited since they require large land areas to guarantee a good quality treatment (Kivaisi, 2001; Ghosh and Gopal, 2010; Foladori et al., 2013). This area can even be enlarged if different CW stages are necessary (Foladori et al., 2012), e.g. a first stage that provides aerobic conditions focusing on organic matter removal/nitrification and a second stage that provides anoxic conditions targeting denitrification. The CW space requirements can become a limiting factor for example in densely populated areas, in mountain regions and in situations when local authorities demand the treatment of wastewater before discharge.

Vertical flow CWs (VF CWs) are generally sized in Europe with 1-3 m^2 PE^{-1} (population equivalent) and horizontal flow CWs (HF CWs) with 5 m^2 PE^{-1} (Vymazal, 2011). The design depends on factors such as effluent needs, ambient temperatures, technology combinations and use of energy. If land area requirement is the main factor that decides the selection of a wastewater treatment system, other technologies such as activated sludge (0.2-0.4 m^2 PE^{-1}), trickling filters (0.3-0.7 m^2 PE^{-1}) or upflow anaerobic sludge blanket reactors (0.05-0.4 m^2 PE^{-1}) (von Sperling, 1996; Mburu et al., 2013) can become the foremost option. Since CWs are natural treatment technologies that at the same time provide green areas, it is important to design CWs capable of appropriate wastewater treatment while assuring a smaller footprint.

Thus, this study aimed to develop a CW setup, called Duplex-CW, to be used when land availability is scarce. A Duplex-CW is a hybrid system that combines two compartments in a stacked design: a VF CW on top of a horizontal flow filter (HFF), similar to the system developed by Kantawanichkul et al. (2001). The Duplex-CW was partially studied in Chapter 5, but further research remains necessary (e.g. capability of area reduction). In consequence, the specific design of the Duplex-CW is yet not defined and therefore the objectives of this research were: (i) to assess the differences among three different Duplex-CW configurations subjected to different domestic wastewater strengths, (ii) to select the most appropriate configuration for the Duplex-CW that can reduce the area requirements without deteriorating the effluent quality and (iii) to evaluate the need of (intermittent) artificial aeration in the Duplex-CW design.

6.2. MATERIALS AND METHODS

6.2.1. Experimental setup

Three laboratory scale Duplex-CWs, planted with *Phragmites australis*, were evaluated in this study. The support medium was coarse sand (1-2 mm) and the drainage layer consisted of gravel (15-30 mm). Each Duplex-CW had a surface area of 0.24 m^2, while the depths were 0.80 m (0.70 m of sand and 0.10 m of drainage layer) for the VF CW and 0.35 m (only sand) for the HFF (Fig. 6.1, Appendix E). To provide artificial aeration to the VF CWs, perforated horizontal pipes were placed between the sand and gravel layer (Appendix E). The systems were operated in a greenhouse under controlled temperature (20-23°C) and light intensity (85-100 µmol photons m^{-2} sec^{-1} for 16 h d^{-1}).

The wastewater was applied intermittently, with a peristaltic pump, on top of the VF CW by means of a pipe manifold, twice per week (three batches of 13 L each day, batch interval of 6 h) corresponding to a hydraulic loading rate (HLR) of ~0.046 m^3 m^{-2} d^{-1}. The wastewater used was primary effluent from Harnaschpolder domestic wastewater treatment plant (Delft, The Netherlands) that was allowed to settle for approximately 2 h before its use. The physical and chemical characteristics of the settled wastewater are given in Table 6.1. This wastewater was applied during a 2-months start-up/adaptation period (previous to the experiments).

The three configurations of the Duplex-CW were named fill and drain (Fill&D), stagnant batch (StagB) and free drain (FreeD), following the different functioning modes of their VF CWs (Fig. 6.2). In the Fill&D system, three batches of wastewater were added while the outlet valve was closed. After 1 d, the valve was opened and water drained into the HFF (Fig. 6.2A). In the StagB system, an elbow joint (17 cm height) was installed at the outlet of the VF CW to retain 1.25 batches (16.25 L) of wastewater (stagnant wastewater) (Fig. 6.2B, Appendix E). The time between two consecutive batches was ~6 h within a feeding day and 3-4 d between the last batch and the first batch of two consecutive feeding days, therefore the HRT in this configuration varied between 6 h and 4 d (Fig. 6.2B). In the FreeD system, the outlet (valve) of the VF CW was permanently open enabling the water to directly discharge to the HFF in ~1.5 h (Fig. 6.2C). The HFF of all configurations worked similarly and had a HRT of 3-4 d.

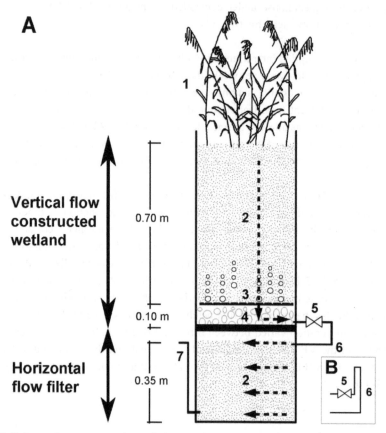

Figure 6.1 Schematic representation of the Duplex-constructed wetland configurations (A) used in this study. 1- *Phragmites australis*, 2- Sand (support media), 3- Aeration pipe, 4- Gravel (drainage layer), 5- Valve, 6- Pipe connecting both compartments and 7- Outlet pipe. The dashed lines show the path of the wastewater in the system. The graph on the bottom-right (B) represents the modification (elbow), done for the "Stagnant batch" configuration (also found in Appendix E).

Table 6.1 Composition of the primary settled domestic wastewater used in this study (n = 9).

Parameters	Unit	Mean ± Std. dev.
pH	-	7.0 ± 0.1
Electrical conductivity	µS cm^{-1}	1271 ± 175.3
Dissolved oxygen	mg L^{-1}	1.0 ± 0.6
Chemical oxygen demand	mg L^{-1}	329 ± 87.2
Total suspended solids	mg L^{-1}	118 ± 21
NH$_4^+$-N	mg L^{-1}	43 ± 7.5
NO$_3^-$-N	mg L^{-1}	0.1 ± 0.1
Total nitrogen	mg L^{-1}	47 ± 9.5
Total phosphorus	mg L^{-1}	9.0 ±1.0

The variations in the operational characteristics of each VF CW were done with the intention of creating different oxygen conditions: (i) Fill&D, the resting period in between feeding days assured an aerobic bed that facilitated aerobic processes when the wastewater was introduced; (ii) StagB, the permanent saturated bottom layer (stagnant batch) and the unsaturated top layer kept within the VF CW, created both anoxic-anaerobic and aerobic zones, and (iii) FreeD, the wastewater trickling along the depth assured permanent aerobic conditions in the VF CW bed.

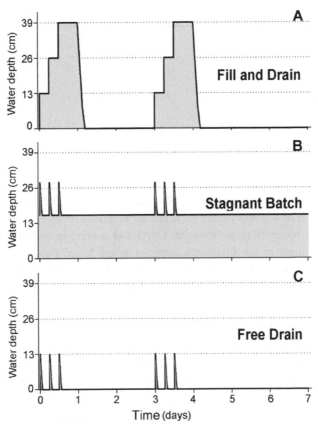

Figure 6.2 Weekly hydraulic behaviour of the three Duplex-constructed wetland (CW) configurations used in this study. Each batch of wastewater contained 13 L and had a depth of 13 cm in the vertical flow CW (VF CW).

6.2.2. Experimental design

After 2 months for the start-up of the systems, four experimental periods were tested in all Duplex-CWs (Table 6.2). In the first three periods, the performance of the Duplex-CW configurations was tested using three different domestic wastewater strengths for a total period of 17 weeks (Table 6.2). For acclimation, a feeding day with the new type of wastewater was provided prior to starting a new experiment. From here onwards, the wastewater types will be indicated with WW, WW^+ and WW^{++} from low to high strength wastewater, respectively. The WW type consisted of the primary settled wastewater (~330 mg COD L^{-1}, Table 6.1); for WW^+ and WW^{++}, peptone was added to increase the strength (mainly of organic matter and nitrogen) and reach COD concentrations of ~600 (0.3 g peptone L^{-1}) and ~800 (0.5 g peptone L^{-1}) mg L^{-1}, respectively.

Table 6.2 Characteristics of the influent used in the different periods of this study.

Period	Experiment	Abbreviation	COD (mg L^{-1})	Organic loading rate (g COD m^{-2} d^{-1})	Experimental weeks	Area* (m^2 PE^{-1})
1	Wastewater	WW	330	15	9	7.9
2	Wastewater + 0.3 g peptone L^{-1}	WW^+	600	27	4	3.4
3	Wastewater + 0.5 g peptone L^{-1}	WW^{++}	800	37	4	2.6
4	Wastewater + 0.5 g peptone L^{-1} + aeration	WW_A^{++}	800	37	4	2.6

*The population equivalent (PE) was calculated using the common relation 1 PE = 60 g BOD_5 d^{-1}. BOD_5 values were approximated using the following ratio COD/BOD_5 = 2.

In the fourth experimental period, artificial aeration was applied to all VF CWs fed with WW^{++} for a period of 4 weeks (Table 6.2). The air flow was set to ~2 L min^{-1} (Appendix B) using an air flow meter (Key Instruments, USA) and applied in an intermittent mode that started at the moment of the first batch application and lasted for 24 h. A picture of the aeration pipes is shown in Appendix E. The results of this experiment (WW^{++} with aeration, WW_A^{++}) were compared with the results obtained with WW^{++} (without aeration).

6.2.3. Sample collection and analytical methods

Wastewater samples were collected weekly at the inlet and outlet of each compartment of the Duplex-CW during the four experimental periods and analyzed according to the procedure outlined in APHA (2005): for pH and dissolved oxygen (DO) by the electrometric method, COD by the open reflux titrimetric method, TSS by the gravimetric method, total nitrogen (TN) and total phosphorus (TP) digestion by the persulfate method followed by measurements of NO_3^--N (ultraviolet spectrophotometric screening method) and PO_4^{3-}-P (vanadomolybdophosphoric acid colorimetric method), respectively. The NH_4^+-N concentration was measured by the dichloroisocyanurate method according to Dutch Standards (NEN 6472, 1983) and, NO_3^--N by ion chromatography (ICS-1100, DIONEX™, USA). The samples at the outlet of the VF CW were the same as the inlet of the HFF.

Organic matter compounds (i.e. humic-, fulvic- and protein-like) were analyzed by measuring the fluorescence excitation emission matrix (EEM) spectra of samples from the WW^+, WW^{++} and WW_A^{++} periods (n = 1) as described by Zapater-Pereyra et al. (2014) using MATLAB (version R2012b) to identify the compounds in contour maps as peaks of an EMM after correction of the intensities with the DOC dilution factor and after subtraction of the EEM of a blank (Milli-Q water).

When the experimental periods were finalized, the diffusion of oxygen in the VF CWs was quantified by monitoring the effluent DO and temperature of the anoxic water that was added. Fill&D and StagB CWs were emptied before the addition of anoxic water. The anoxic water was prepared in a container by mixing 1.4 g of Na_2SO_3 (as oxygen scavenger) and 33 mg of $CoCl_2$ (as catalyst) with demineralized water (20 L, 20.9-22°C) and flushing it with nitrogen gas. To add the anoxic water to the VF CW, the tap of the container was connected to the pipe manifold of the VF CW with a butyl tubing. Samples were collected from the outlet of the VF CWs at time zero and at intervals of 10 min for a period of 2 h.

Above-ground biomass was harvested 3 times for each configuration (after sequential periods of 151, 98 and 105 d, n = 3) and only ~5 cm of stems were left. The harvested biomass (leaves and stems) were mixed, a sub-sample was separated to measure nutrient content and the rest was dried (105 °C for at least 48 h) to measure the total biomass content (weight). After 435 d (1.2 yr) of operation (~5.7 months after the WW_A^{++} experiment finished) the setups were dismantled to quantify the total below-ground dry biomass content (70°C for at least 48 h, n = 1) and nutrient content. The type of wastewater received after the WW_A^{++} experiment was WW^{++} for 3.0 months and WW for 2.7 months. Nutrient content (TN and TP) in the dry media (70 °C) was analyzed by digestion with H_2SO_4/Se/salicylic acid and H_2O_2 (Walinga et al., 1989). Digestion was followed by NH_4^+-N and PO_4^{3-}-P analysis by the colorimetric method (NEN 6472, 1983) and the ascorbic acid spectrophotometric method (APHA, 2005), respectively.

When dismantling the systems, a large sand sample was taken at depths of 0-20, 21-40, 41-60 and 61-70 cm from the VF CWs and at the inlet, middle and outlet from the HFFs. Each sample was properly homogenized and 3 sub-samples (15 mL each ≈ 21 g dry weight, n = 3) were collected to measure the accumulated solids on the sand. Briefly, each sand sub-sample was mixed with water and sonicated (Soniprep 150, MSE, UK) for 6 min at amplitude of 30 μm. The supernatant (containing the accumulated solids) was filtered with GF/C filters (Whatman^{TM}, UK). The water addition, sonication and filtration were done three times per sub-sample, to assure the removal of all accumulated solids. All filters were dried at 105 °C for 24 h and the accumulated solids on each filter were calculated as in the TSS method. The sum of the solids on the three filters per subsample was reported as the total amount of accumulated solids.

6.2.4. Data analysis

Analysis of variance (ANOVA) followed by Tukey post-hoc test for all pairwise multiple comparison (α = 0.05) were used to compare: (i) the differences between the wastewater strengths in each Duplex-CW configuration compartment per parameter measured, (ii) the differences between the configurations' solid accumulation per compartment (combining all depth dta for the VF CWs (n = 12) and length data for the HFFs (n = 9)), (iii) the differences between configurations of Duplex-CW above-ground biomass nutrient concentration per parameter measured (TN and TP) and (v) the differences between the configurations' above-ground biomass dry weight. Normality assumption and equal variance were tested using the Shapiro-Wilk and Levene median test, respectively. If assumptions were not met, values were $\log_{10}$-transformed. If the transformation was not useful to meet assumptions, ANOVA on ranks (Kruskal-Wallis) followed by the pairwise comparison using Dunn's method was conducted. Comparison between the aerated and non-aerated Duplex-CWs was done by applying a two tailed paired T-test ($\alpha = 0.05$). If the normality assumption was not met, the Wilcoxon signed rank test was performed. All tests were conducted using SigmaPlot 12.3 software.

6.3. RESULTS

6.3.1. Influence of different domestic wastewater strengths on the performance of the VF CW and HFF compartments

WW. The influent WW had concentrations of 330 mg L^{-1} for COD, 120 mg L^{-1} for TSS, 43 mg L^{-1} for NH_4^+-N, 47 mg L^{-1} for TN and 9 mg L^{-1} for TP (Table 6.1). All configurations tested reached similar removal efficiencies (Table 6.3) with final effluent concentrations of the same order of magnitude (Figs. 6.3 and 6.4). Only NH_4^+-N achieved a better removal in the Fill&D (6 mg L^{-1}) as compared to the other two configurations (12-13 mg L^{-1}, Fig. 6.4).

In the Fill&D configuration, the major treatment location (from this point forward, the "major treatment location" refers to the compartment - VF CW or HFF - that provided most of the treatment) was given by the VF CW. The Fill&D VF CW contributed the most to the removal of all the other parameters (except for TN, Table 6.3) and to the production of NO_3^--N (from 0 to 17 mg L^{-1}, Fig. 6.4A). The VF CW from the StagB configuration as well contributed the most to the removal of almost all parameters, except for TP (Table 6.3). In this system there was no increment of NO_3^--N in any compartment (Fig. 6.4B), despite that the NH_4^+-N and TN concentrations decreased (Fig. 6.4E and H). On the contrary, the HFF compartment was more active in the FreeD configuration (Table 6.3), except for NH_4^+-N and NO_3^--N that were highly removed and produced, respectively, in the VF CW.

Table 6.3 Removal efficiency and compartment where the majority of the treatment occurred in the Duplex-constructed wetland. Sample size of $n = 9$ for the WW period and $n = 4$ for each of the other periods.

	WW			WW+			WW++			WW$_A^{++}$		
	Total %rem.	Major treatment*		Total %rem.	Major treatment*		Total %rem.	Major treatment*		Total %rem.	Major treatment*	
		Location	%rem.		Location	% rem.		Location	%rem.		Location	% rem.
Fill and drain												
COD	87	VF CW	69	93	VF CW	65	91	VF CW	70	95	VF CW	61
TSS	91	VF CW	69	93	VF CW	55	84	VF CW	70	89	VF CW	62
NH$_4^+$-N	85	VF CW	82	73	VF CW	51	55	VF CW	46	72	VF CW	58
Total Nitrogen	72	HFF	38	82	HFF	49	78	HFF	48	71	HFF	43
Total Phosphorus	80	VFCW	42	61	VF CW	50	44	VF CW	-	66	VF CW	59
Stagnant batch												
COD	88	VF CW	-	93	VF CW	52	87	VF CW	62	93	VF CW	75
TSS	94	VF CW	50	84	VF CW	45	64	VF CW	53	83	VF CW	67
NH$_4^+$-N	72	VF CW	-	42	VF CW	30	15	HFF	-	44	VF CW	27
Total Nitrogen	70	VF CW	-	61	VF CW	47	61	VF CW	49	52	VF CW	41
Total Phosphorus	81	HFF	46	58	HFF	46	52	VF CW	33	74	VF CW	39
Free drain												
COD	89	HFF	53	85	VF CW	44	71	HFF	41	87	HFF	48
TSS	92	HFF	57	76	VF CW	39	50	VF CW	39	74	HFF	42
NH$_4^+$-N	71	VF CW	46	43	VF CW	40	12	VF CW	-	75	VF CW	42
Total Nitrogen	60	HFF	57	60	HFF	45	46	HFF	32	52	HFF	48
Total Phosphorus	76	HFF	48	39	VF CW	-	27	VF CW	-	63	VF CW	37

* Major treatment "Location" and "% rem." refer to the compartment that provided the majority of the total removal (vertical flow constructed wetland (VF CW) or horizontal flow filter (HFF)) and to the percentage removal that such compartment achieved out of the "Total % removal", respectively. For example for Fill and drain, WW strength, COD: 69% of the total COD removed (87%) was removed by the VF CW, the remaining (18%) was removed by the HFF. Therefore the VF CW is the compartment providing the majority of the removal.

The Major treatment % rem. value was not indicated if the outflow concentration of the compartment in "Location" was larger than the inflow concentration (in order to avoid confusions with negative values).

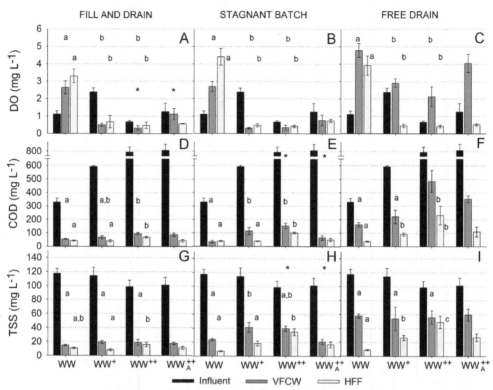

Figure 6.3 Mean (± Std. error) dissolved oxygen, organic matter and solids concentrations of influent and effluent of each Duplex-constructed wetland (CW) compartment (vertical flow - VF CW and horizontal flow filter - HFF) during each experimental period using different domestic wastewater (WW) strengths: WW; WW + 0.3 g peptone L^{-1}, WW^+; WW + 0.5 g peptone L^{-1}, WW^{++} and WW^{++} with aeration, WW_A^{++}.

Statistics note: Letters a,b,c for a certain parameter at a certain compartment within a particular Duplex-CW configuration indicates significant differences between the WW, WW^+ and WW^{++}. Upper and lower letters in each graph are the statistical results of VF CW and HFF, respectively. The symbol * displayed indicate statistical differences for a certain parameter at a certain compartment within a particular Duplex-CW configuration when artificial aeration was applied (WW_A^{++}) or not (WW^{++}).

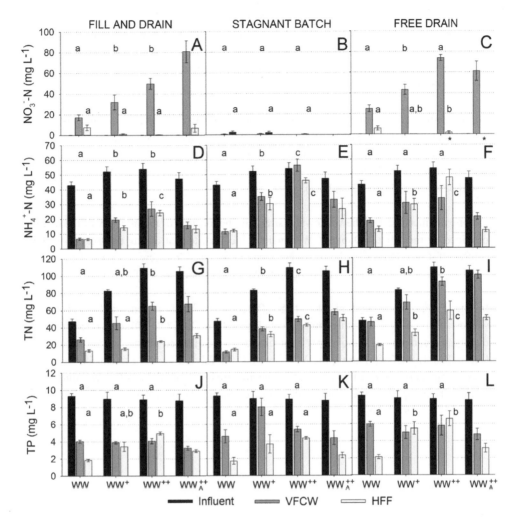

Figure 6.4 Mean (± Std. error) nutrient concentrations of influent and effluent of each Duplex-constructed wetland (CW) compartment (vertical flow - VF CW and horizontal flow filter - HFF) during each experimental period using different domestic wastewater (WW) strengths: WW; WW + 0.3 g peptone L^{-1}, WW^+; WW + 0.5 g peptone L^{-1}, WW^{++} and WW^{++} with aeration, WW_A^{++}.

Statistics note: Letters a,b,c for a certain parameter at a certain compartment within a particular Duplex-CW configuration indicates significant differences between the WW, WW^+ and WW^{++}. Upper and lower letters in each graph are the statistical results of VF CW and HFF, respectively. The symbol * displayed indicate statistical differences for a certain parameter at a certain compartment within a particular Duplex-CW configuration when artificial aeration was applied (WW_A^{++}) or not (WW^{++}).

WW$^+$ and WW^{++}. The composition of the WW$^+$ and WW^{++} differed from the original WW only for COD and TN, whereas NH$_4^+$-N and NO$_3^-$-N concentrations were similar (Figs. 6.3 and 6.4). Both WW$^+$ and WW^{++} contained, for most of the cases, the three investigated organic matter compounds (humic-, fulvic- and protein-like), and their peak intensities were higher in the WW^{++} (Table 6.4).

For all configurations tested, the increase in wastewater strength deteriorated the COD, NH$_4^+$-N, TN and TP effluent quality of each compartment, while this was not so evident for the NO$_3^-$-N and TSS concentrations. Statistical comparison (per compartment, per configuration) showed significant differences between the three wastewater types for the majority of the tested parameters (Figs. 6.3 and 6.4).

Despite the increment in the wastewater strength, the location of the majority of the treatment provided by the Fill&D Duplex-CW remained identical to that during the treatment of WW (in the VF CW). However, the treatment location reversed in a few cases for the StagB and FreeD systems as compared to that encountered when treating WW (Table 6.3).

In the Fill&D configuration, the TSS effluent concentration remained similar for all wastewater types and at each compartment (Fig. 6.3D). The COD effluent concentration and the organic compounds removal per compartment were affected mainly when treating WW^{++} (Fig. 6.3A, Table 6.4). During the WW$^+$ period, the fulvic-like and protein-like compounds were completely removed (100%) after the treatment of only the VF CW compartment. Humic-like compounds were never fully removed from WW$^+$ and WW^{++}.

Table 6.4 Peak intensity (% reduction of peak intensity from the influent) of humic-like, fulvic-like and protein-like organic matter compounds (n = 1) per compartment (vertical flow constructed wetland (VF CW) and horizontal flow filter (HFF)) of each Duplex-CW when applying WW^+, WW^{++} and WW_A^{++}.

| | WW^+ | | | WW^{++} | | | WW_A^{++} | | |
	Influent	VF CW	HFF	Influent	VF CW	HFF	Influent	VF CW	HFF
Fill and drain									
Humic-like	8.3	4.3 (48)	3.5 (58)	10.9	6.8 (38)	4.9 (55)	14.6	6.6 (55)	5.5 (62)
Fulvic-like	10.2	0.0 (100)	0.0 (100)	12.2	5.1 (58)	3.5 (71)	13.9	5.0 (64)	4.0 (71)
Protein-like	3.3	0.0 (100)	0.0 (100)	21.5	8.1 (62)	5.3 (75)	14.1	6.2 (56)	0.0 (100)
Stagnant batch									
Humic-like	9.2	4.6 (50)	3.3 (64)	12.5	11.5 (8)	4.9 (61)	16.8	6.7 (60)	4.7 (72)
Fulvic-like	ND	ND	ND	8.2	8.0 (2)	0.0 (100)	15.4	4.6 (70)	3.3 (79)
Protein-like	9.5	0.0 (100)	0.0 (100)	27.2	9.8 (64)	0.0 (100)	14.6	0.0 (100)	0.0 (100)
Free drain									
Humic-like	6.9	5.2 (25)	3.6 (48)	11.3	8.6 (24)	7.9 (30)	15.9	7.0 (56)	4.1 (74)
Fulvic-like	ND	ND	ND	12.1	0.0 (100)	9.0 (26)	7.1	6.6 (7)	0.0 (100)
Protein-like	9.5	0.0 (100)	0.0 (100)	28.4	14.7 (48)	8.5 (70)	11.8	0.0 (100)	0.0 (100)

Excitation-Emission wavelength (λEx/Em): 240-260/ 410-450 nm for humic-like, 290-340/ 410-430 nm for fulvic-like and 270-280/ 300-350 nm for protein-like.
ND: Not detected.

The nutrients removal by the Fill&D system was also affected by the increment in the wastewater strength. The TP concentration in the VF CW effluent maintained constant in the three wastewater types tested, but the HFF effluent concentration increased with the increase in strength despite the unchanged influent TP concentration (Fig. 6.4J). For NH_4^+-N and TN (Fig. 6.4D and G), the higher the wastewater strength, the higher the VF CW and HFF effluent concentration. For NO_3^--N (Fig. 6.4A), the same occurred after the VF CW compartment and once in the HFF, this parameter strongly decreased. Moreover, the HFF removed further NO_3^--N and TN (from the VF CW effluent concentration), but not NH_4^+-N.

The use of WW^+ and WW^{++} in the StagB configuration highly affected its performance when removing TSS, NH_4^+-N and TN (Table 6.3) reaching effluent concentrations of more than double than those achieved when treating WW (Figs. 6.3 and 6.4). The overall removal efficiency of COD was not affected (Table 6.3), but the effluent concentrations of each compartment highly increased (Fig. 6.3E). NO_3^--N and TP remained unaffected (p>0.05, Fig. 6.4B and K). Similar to the Fill&D configuration, the peak intensity of the humic-like compounds were not totally removed (up to 64%). However, with this system both fulvic- and protein-like compounds were 100% removed in both cases (WW^+ and WW^{++}). The major treatment location remained identical during the WW^+ period as compared to the WW period. During the WW^{++} period, the NH_4^+-N and TP treatment location reversed to the HFF and VF CW, respectively (Table 6.3).

The removal of all parameters shown in Table 6.3 greatly declined in the FreeD configuration. Its final effluent concentrations using WW^+ and WW^{++} were higher than those from the StagB and Fill&D Duplex-CW for COD, TSS and TP (Figs. 6.3 and 6.4). The NH_4^+-N and TN concentrations were higher than those from the Fill&D, but similar to those from the StagB. The NO_3^--N final effluent concentration was similar in the three configurations and the behaviour in the system was similar to that in the Fill&D configuration (Fig. 6.4C). This configuration was only able to completely remove the protein-like organic compounds when using WW^+. The other organic compounds in both types of wastewater were not totally removed (Table 6.4). The major treatment location shifted in many cases from the HFF to the VF CW, except for NH_4^+-N (the VF CW) and TN (the HFF) that remained unchanged despite the higher wastewater strengths.

6.3.2. Effect of artificial aeration on the treatment of WW^{++}

For almost all the cases, the use of artificial aeration did not provide statistical differences at a certain compartment (p>0.05, Figs. 6.3 and 6.4, presence of * symbol indicates significant difference). No statistics can be conducted for the organic matter compounds due to the sample size (n = 1), however the use of artificial aeration likely enhanced the removal of the organic matter compounds in all configurations investigated (Table 6.4). The protein-like compounds were the only observed peak in all Duplex-CW configurations that was consistently reduced to 100% when aeration was provided (Table 6.4). In contrast, when aeration was not applied, the total removal of protein-like compounds did not always occur.

For all cases, humic-like compounds showed higher peak intensity as compared to the peaks of other compounds, despite the presence or absence of artificial aeration.

No variations were observed in the major treatment location for almost all cases. In other words, contaminants that were removed mainly in the VF CW or HFF during the WW^{++} period, were also removed in that compartment during the WW_A^{+} period, except for NH_4^+-N in the StagB and TSS in the FreeD configuration (Table 6.3).

6.3.3. Solids accumulation on the sand, plant biomass and nutrient uptake

The accumulated solids on the sand (expressed per volume of sand in Fig. 6.5 and by the total mass in Table 6.5) of the Fill&D HFF was lower from that in the VF CW. In the StagB and FreeD systems, the HFF showed higher solids accumulation per volume of sand (2-8 mg_{solids} mL^{-1}_{sand}) than that on the VF CW (1-4 mg_{solids} mL^{-1}_{sand}) (Fig. 6.5). In terms of total mass, the VF CW and HFF from the StagB had a similar solids accumulation (353 and 325 g, respectively) while for the FreeD, the solids accumulated in the HFF (454 g) were much higher than those accumulated in the VF CW (305 g) (Table 6.5). There were significant differences ($p<0.05$) among the VF CW and as well among the HFF compartments of the 3 systems, with the Fill&D system being significantly different than the FreeD setup in both compartments and the StagB setup in only the HFF (Fig. 6.5). The solids accumulation rate of the Fill&D HFF compartment was less than half than in all other compartments (Table 6.5).

Table 6.5 Amount of solids accumulated on the sand and solids accumulation rates in both compartments of the Duplex-constructed wetland configurations at the end of the study.

Duplex-CW	Calculated total accumulated solids* (g)	Measured total accumulated solids (g)	Measured total accumulated solids (kg m^{-2})	Measured total accumulated solids rates (kg m^{-2} yr^{-1})
	VF CW/HFF	VF CW/HFF	VF CW/HFF	VF CW/HFF
Fill and drain	421/18	406/137	1.7/0.6	1.5/0.5
Stagnant batch	357/54	353/325	1.5/1.4	1.3/1.2
Free drain	240/145	305/454	1.3/1.9	1.1/1.7

* Calculated values obtained from the total suspended solids applied with the wastewater considering the average removal values per experimental period per compartment.

For each configuration, the amount of accumulated solids expected to be on the sand was calculated using the amount of TSS in and out each compartment and was compared to that measured (Table 6.5). For the VF CW compartments, the calculated accumulated solids were in the same order of magnitude to the values measured. On the contrary, the measured values for HFFs were much higher (3-7 times more) than those provided by the TSS load (Table 6.5).

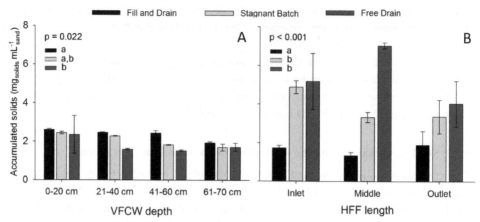

Figure 6.5 Mean (± Std. error) solids accumulation on the sand in the vertical flow constructed wetland (VF CW) and horizontal flow filter (HFF) of each Duplex-CW configuration. The p-value in each graph (A and B) show the statistical significance among the three configurations and the a,b letter indicate their pairwise multiple comparison along the depth (A) and length (B).

The above-ground biomass and nutrient (nitrogen and phosphorus) uptake was similar for all configurations tested (p>0.05, Table 6.6). The above-ground biomass removed 3-5% of TN and 4-5% of TP from the influent. Below-ground biomass and nutrient uptake showed slight differences among the 3 configurations; however values were in the same order of magnitude (Table 6.6). The nutrient uptake (from the influent) contribution by below-ground biomass was 0.4-1.0%.

Table 6.6 Above- and below-ground biomass quantification (dry weight) and nutrient uptake in the Duplex-constructed wetland configurations.

	Dry weight (above-ground: g d^{-1})* (below-ground: g)	Total nitrogen (mg g$_{dry\ weight}$$^{-1}$)	Total nitrogen (g m^{-2})	% rem.	Total phosphorus (mg g$_{dry\ weight}$$^{-1}$)	Total phosphorus (g m^{-2})	% rem.
Above-ground (Mean ± Std. error, n = 3)							
Fill and drain	1.6 ± 0.1^a	23.2 ± 2.1^a	67	5	3.2 ± 0.1^a	9	5
Stagnant batch	1.3 ± 0.3^a	24.1 ± 1.5^a	57	4	3.7 ± 0.3^a	9	5
Free drain	1.0 ± 0.0^a	22.5 ± 0.5^a	41	3	4.0 ± 0.3^a	7	4
Below-ground (n = 1)							
Fill and drain	106	11.2	4.9	0.4	2.5	1.1	0.6
Stagnant batch	93	14.9	5.8	0.4	3.1	1.2	0.7
Free drain	96	14.9	6.0	0.4	4.7	1.9	1.0

Superscripts showing only the letter "a" indicate that there were no statistical differences among the Duplex-CW configurations, per parameter.

* Values are expressed in g d^{-1} as each period between harvesting of the above-ground biomass lasted slightly different (i.e. 151, 98 and 105 d). To obtain the dry biomass in g m^{-2}: dry weight (in g d^{-1}) × operational period (in d) / area (0.24 m^2). The total operational period of the systems was 435 d.

6.3.4. **VF CW oxygen diffusion experiment**

After the addition of the anoxic water (DO_{mean} = 0.9 mg L^{-1}) to the 3 VF CWs, the DO levels started to increase to > 5 mg L^{-1}. The FreeD reached its maximum DO value in 10 min and then maintained at similar levels for 1.5 h, while the Fill&D and StagB VF CWs required 20 min to reach the maximum DO value and immediately decreased back to levels of ~3 mg O_2 L^{-1} (Fig. 6.6).

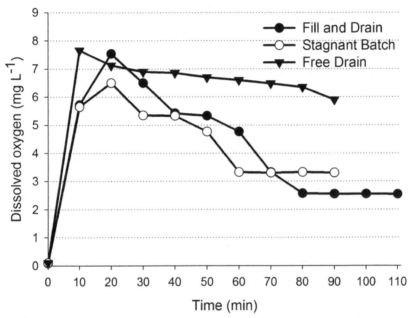

Figure 6.6 Dissolved oxygen concentrations as a function of operational time in the three vertical flow constructed wetland (CW) compartments of the Duplex-CW configurations during the oxygen diffusion experiment.

6.4. DISCUSSION

6.4.1. **Wastewater strength**

Enriching the wastewater with peptone highly increased the initial COD concentration (from ~330 to ~600-800 mg L^{-1}) but did not create additional solids (Fig. 6.3). The similar TSS and COD effluent concentration trend, per configuration, suggested that the added dissolved organic matter was degraded first (easily biodegradable) and the particulate organic matter remained (more resistant to biodegradation) in the Duplex-CWs. The relatively high DO concentration in the effluent of each compartment (> 2 mg L^{-1}, Fig. 6.3) during the WW period suggested that the remaining oxic environment in the systems could be capable of treating more polluted wastewater. But, treating enriched wastewater lowered the DO to anoxic levels incapable to provide aerobic degradation as during the WW period (Fig. 6.3A-C). This resulted in the deterioration of the COD concentration with the increase of strength, as aerobic COD degradation is more important in CWs than anaerobic degradation.

Nitrogen removal was also affected by the use of enriched wastewater (Table 3). Higher NH_4^+-N effluent concentrations were found in all configurations when treating WW^+ and WW^{++} as compared to the WW period (Fig. 6.4D vs. E-F) because the available DO was consumed in the degradation of COD (Fig. 6.3), thus not enough DO was available for nitrification. Moreover, the use of peptone also affected the NH_4^+-N removal. Peptone is a rich source of organic nitrogen; hence, it increased the TN influent concentration but not the inorganic nitrogen concentration (Fig. 6.4). It should be noted that in common domestic wastewater 35-40% of TN is organic nitrogen (Metcalf and Eddy, 2003), similar to this study (37% for WW^+ and 50% for WW^{++}). This organic nitrogen was quickly and completely (100%) converted to NH_4^+-N by ammonification (Vymazal, 2007), thus increasing the load of NH_4^+-N to the systems. Ammonification was indirectly verified since the sum of inorganic nitrogen effluent concentrations was similar to that of TN for each compartment.

The organic compounds introduced by the use of peptone were protein-like compounds (Table 6.4), as peptone is made up of chains of amino acids. Hence, changes in the influent concentration of other compounds (carbohydrates and fats) are solely the quality fluctuations of the wastewater. Fulvic- and protein-like organic compounds from WW^+ were completely removed (100%) in all configurations, while from WW^{++}, only the StagB configuration (after the HFF) was able to completely remove both organic compound peaks (Table 6.4). None could remove humic-like compounds further than 64% (Table 6.4). It has been reported that the hydrophobicity of humic substances enhances their resistance to biodegradation (i.e. recalcitrant) (Dignac et al., 2000), while fulvic- (Saar and Weber, 1982) and protein-like (Nam and Amy, 2008) organic compounds are strongly hydrophilic, thus more prone to degradation (i.e. labile). Furthermore, their molecular weight also plays a role in their removal. According to Imai et al. (2002), the molecular weight of a hydrophilic fraction of humic substances is by far smaller than that of the hydrophobic fraction. For instance, Dignac et al. (2000) observed that proteins were the second most removed compound after lipids from wastewater treated in an activated sludge system.

In this study, the StagB was the most anoxic configuration (< 1 mg L^{-1} in both compartments for WW^+ and WW^{++}, Fig. 6.3) and had the lowest oxygen diffusion (Fig. 6.6). Nevertheless, it showed a better treatment performance of all organic compounds suggesting that adsorption was an important removal mechanism of organic matter in this configuration, and plausibly in the others as well.

Overall, the Fill&D system was the less affected configuration by the enriched wastewater due to the longer HRT in the VF CW as compared to that in the other systems (Ghosh and Gopal, 2010; Weerakoon et al., 2013). Nevertheless, during the first 3 wastewater periods, the effluent concentrations and removal efficiencies of all Duplex-CW configurations are comparable to those achieved by the 19 conventional hybrid CWs (VF-HF sequence) reviewed by Vymazal (2013) treating similar wastewater strengths.

6.4.2. Contribution of the VF CW and HFF compartments to pollutant removal

The VF CW compartment of the Fill&D and StagB provided the majority of the treatment for organic matter and solids, in agreement with the results found by Kantawanichkul et al. (2001) in a similar system and with those found in Chapter 5 using a Fill&D configuration (Fig. 5.2B-D). The NH_4^+-N concentrations in all systems were mainly removed in the VF CW compartment and, for all wastewater periods, Fill&D showed lower NH_4^+-N concentrations than the other two systems, as reported in Chapter 5 (Sections 5.3.3 and 5.4.2). The Fill&D and StagB had a longer HRT (> 3 h) as compared to the FreeD system (~1.5 h). Longer HRTs usually result in better pollutant removal efficiencies (Ghosh and Gopal, 2010; Weerakoon et al., 2013). The FreeD system had the higher DO concentration (in all wastewater periods, Fig. 6.3) and the best water passive oxygenation (Fig. 6.6). The continuous water movement/drainage towards the HFF (drain causes suction of fresh air) and the long resting periods reoxygenate this compartment fast. But this compartment was highly affected by the short HRT. The ~1.5 h was not enough to provide an appropriate treatment and therefore, biodegradation occurred in the HFF first aerobically (as it came from the VF CW) and then anaerobic-anoxically (when oxygen was depleted) (Fig. 6.3C). Literature reports experiences with HRTs of above 10 d for an effective treatment (Kadlec and Wallace, 2009; Ghosh and Gopal, 2010). Organic matter removal (biochemical oxygen demand, BOD_5) is critical below 1 d HRT and improves until a HRT of 7.5 d (Reed and Brown, 1995 cited in Weerakoon et al., 2013). The FreeD was thus in the critical range to provide sufficient organic matter removal.

The combination of saturated and unsaturated zones in the StagB VF CW contributed to the simultaneous production and elimination of NO_3^--N within the same compartment. In other words, it contributed to the TN removal without the need of a HFF. The HFF compartment was mainly necessary after the more aerobic VF CWs (Fill&D and FreeD). The absence of a saturated layer in the VF CW of the Fill&D and FreeD did not provide sufficient anoxic conditions for denitrification.

The compartment contribution is as well highly influenced by their position in the Duplex-CW design. As the VF CW is the first compartment, it posses more chances for a higher removal of pollutants. Probably, if the design would have started with a HFF, that would have been the compartment providing the major treatment. However, the design was intended to have a VF CW at the beginning not only for nitrification but, since it is fed intermittently, it would as well promote (in the resting periods) the mineralization of any accumulated solids due to the aerobic environment (Kadlec and Wallace, 2009; Molle, 2014).

Despite one compartment provided the majority of the Duplex-CW removal (for a certain parameter at a certain configuration), the use of the two compartments is still recommended since the second compartment acts as a buffer when the first compartment cannot cope with a pollutant load (e.g. due to a sudden wastewater strength increment) (Foladori et al., 2012). This effect was not only visualized in the parameters concentration (Figs. 6.3 and 6.4) but in the solids accumulation as well (Fig. 6.5, Table 6.5).

Hence, it is important to maintain both compartments in the Duplex-CW design. However, for practical reasons, a Duplex-CW should aim for a HFF supporting the VF CW treatment rather than providing the major treatment. This is recommended as its position in the Duplex-CW is at a complex location (bottom) to provide maintenance (e.g. due to clogging) if needed. Therefore, high solids accumulation in the HFF media should be avoided, placing the FreeD Duplex-CW in disadvantage compared to the other systems.

6.4.3. Aeration

A close look at the effluent concentration of all parameters suggests that the use of aeration for the treatment of WW^{++} improved the water quality (Figs. 6.3 and 6.4, Table 6.4). The few statistical differences suggest that the use of artificial aeration did not provide extra benefits to the Duplex-CWs performance (Figs. 6.3 and 6.4, * symbol). Many studies have claimed that the use of artificial aeration enhances the removal of many pollutants namely COD, NH_4^+-N and TN (e.g. Dong et al., 2012; Hu et al., 2012; Fan et al., 2013c; Foladori et al., 2013; Zapater-Pereyra et al., 2014 - Chapter 4) (Table 5.4). However, some studies are conducted in HF CWs where the aeration effect probably plays a major role since oxygen transfer in HF CWs is lower than that in VF CWs (e.g. Fan et al., 2013c; Zapater-Pereyra et al., 2014 - Chapter 4). Others (e.g. Dong et al., 2012) mention the benefits of aeration (without supporting it with statistics) when the actual concentration differences (between aerated and non aerated) are not tremendously different.

For "gently" artificially aerated systems, Kadlec and Wallace (2009) referred to an oxygen delivery of 50-100 g O_2 m^{-2} d^{-1}. In this study, the oxygen demand to treat the WW^{++} was 30 g O_2 m^{-2} d^{-1}. The average oxygen transfer rate of 22 VF CWs given by Kadlec and Wallace (2009) is 3.5-24.7 g O_2 m^{-2} d^{-1}. In other words, the oxygen demand of the Duplex-CWs when treating WW^{++} was very close to the oxygen transfer rate provided by non aerated VF CWs. Thus, aeration was not strictly necessary. Delivering 100 g O_2 m^{-2} d^{-1} to the Duplex-CWs would imply that the Duplex-CW systems could cope with ~4000 mg COD L^{-1} (186 g BOD_5 m^{-2} d^{-1}). It should not be immediately assumed, with this calculation, that adding more air to the Duplex-CW would treat completely any concentration of organic matter added. But mainly, this relates to an under loaded system yet not capable to fully trigger the benefits of artificial aeration, in agreement with the results found previously in this study (Section 5.4.1). For instance, in the case of nitrogen, Hu et al. (2012) mentioned that "when dealing with high strength wastewater, artificial aeration seems to be the only option to achieve complete nitrification". Another possibility for the little benefit provided by aeration was the poor efficiency of the air bubbles to flow evenly through the sand media (Section 5.4.1, Appendix E).

Although it was not possible to monitor the accumulation of solids with and without aeration in this study, it is important to note that aerobic conditions enhance faster mineralization of accumulated organic matter in the media (Kadlec and Wallace, 2009). Therefore, benefits of aeration should not only be considered on a short time basis (effluent quality), but as well as a long-term benefit (less solid accumulation and clogging). Furthermore, Kadlec and Wallace (2009) pointed out that from the economic perspective, the use of artificial aeration is worth only when the aeration cost is lower than the reduction in capital cost (e.g. aerated CWs occupy less area, thus lower construction costs than non aerated CWs). In that regard, intermittent aeration, as used in this study (24 h from the feeding time), is more recommended than continuous aeration (Dong et al., 2012).

6.4.4. Solids accumulation on the sand

The approximately TSS loading rate applied was 5.1 g TSS m^{-2} d^{-1} and the solids accumulation rates for the VF CWs (1.1-1.5 kg m^{-2} yr^{-1}) and HFFs (0.5-1.7 kg m^{-2} yr^{-1}) (Table 6.5) were within the lower range reported in other studies (0.6-14.3 kg m^{-2} yr^{-1}; e.g. Caselles-Osorio et al., 2007) and lower than in Chazarenc et al. (2009) (7.2-13.2 kg m^{-2} yr^{-1}). However, those systems were operated over 2 yr and the Duplex-CWs in this study for only 1.2 yr.

The operational characteristics of the configurations tested in this study were reflected in the solids accumulated. The short retention time in the FreeD VF CW, as compared to the other configurations investigated, was not enough to allow appropriate sedimentation and filtration of the TSS, hence the solids were transferred to the HFF in a higher amount than in the other configurations.

For the VF CW compartments, the calculated accumulated solids from the TSS applied (in g) were in the same order of magnitude of the values measured (Table 6.5). This suggested that solid mineralization was significant; otherwise the values would have been much higher since the accumulated solids are formed not only with the added TSS but also with biofilm, precipitates and plant litter (Kadlec and Wallace, 2009).

On the contrary, the measured values for the HFF were much higher (3-7 times more) than those calculated from the TSS load (Table 6.5). The HFF were not planted, therefore plant litter cannot contribute to the accumulated solids. Therefore, despite some literature mentioning that biofilms are not a relevant cause of solids accumulation (e.g. Langergraber et al., 2003), the biofilm played an important role in solids accumulation in the HFF (Baveye et al., 1998; Thullner et al., 2002). Zhao et al. (2009) added glucose and starch to enrich the wastewater fed to different VF CWs as a source of soluble and particulate organic substrate, respectively. The glucose- and starch-fed systems were used to investigate the clogging process due to biofilm growth and to the combination of particle accumulation and biofilm growth, respectively. Their glucose-fed systems had more accumulated organic matter as compared to the starch-fed systems, suggesting that biofilm growth governed the solids accumulation. Some studies mentioned that biomass occupy 3-8.5% of the initial soil pore volume (Seifert and Engesgaard, 2007). Furthermore, in saturated soils (i.e. low DO environments) the biodegradation of organic matter is much lower than in aerobic environments while the biomass production (e.g. polysaccharides) continues. It could have also been that some solids accumulated in the VF CW were resuspended in the solution due to abrasion, e.g. caused by the artificial aeration (Zapater-Pereyra et al., 2014 - Chapter 4), that were then trapped in the HFF.

6.4.5. Plant biomass and nutrient uptake

It was expected that the difference in operation of the VF CWs would have an impact on the plants. However, no clear optimal water depth for the growth of *Phragmites australis* is reported in literature (Engloner, 2009). Hence, this can explain the similar biomass weight (above- and below-ground) and nutrient uptake between all systems (Table 6.6).

The nutrients taken by the plants (0.4-5%), above- and below-ground, were a minor contribution to the nutrient removal in the Duplex-CW as commonly occurring in CWs (Kadlec and Wallace, 2009). Nutrient uptake by plants varies depending on the type of plant, climate and growing stage (Wu et al., 2013). An ample range of nutrient uptake (rates) has been reported by many authors: 77-218 mg N m^{-2} d^{-1} (Wu et al., 2013), 143-2304 mg N m^{-2} d^{-1} (Tanner et al., 2005), 1.4-44.4 mg P m^{-2} d^{-1} (as reviewed in García et al., 2010), 48.6 g N m^{-2} and 28.91 g P m^{-2} (in *Pragmites australis*, About-Elela and Hellal, 2012), 0.6-250 g N m^{-2} and 0.01-45 g P m^{-2} (as reviewed in Vymazal, 2007), 583 g N m^{-2} and 62 g P m^{-2} (Lee et al., 2013). The nutrient uptake of above-ground biomass (41-67 g N m^{-2} and 7-9 g P m^{-2}, Table 6.6) and consequent removal rates (94-155 mg N m^{-2} d^{-1} and 17-21 mg P m^{-2} d^{-1}) were within the range of those reported values.

6.4.6. Duplex-CW footprint reduction and design selection

The application of WW, WW^{+} and WW^{++} resulted in a design of 7.9, 3.4 and 2.6 m^2 PE^{-1} (Table 6.2), respectively. Following the European disposal guidelines for effluent concentrations (35 mg L^{-1} for TSS, 25 mg L^{-1} for BOD, 125 mg L^{-1} for COD, 15 mg L^{-1} for TN and 2 mg L^{-1} for TP; EEC, 1991), it is evident that none of the configurations tested in this study can treat TP to the recommended limits. If further TP removal is required to meet disposal guidelines, a phosphorus removal post-treatment is recommended (for more information see Chapter 3). Only the Fill&D system met all the other parameters up to WW^{+} (3.4 m^2 PE^{-1}). The other two configurations failed meeting the TN limit for even the WW period.

Foladori et al. (2013) used a VF CW with recirculation and aeration to reduce the area from 3.6 m^2 PE^{-1} to 1.5 m^2 PE^{-1}. Their local guidelines considered 125 mg L^{-1} for COD, 35 mg L^{-1} for TSS and 70% removal efficiency for TN. When applying these specifications to the Duplex-CWs tested, the Fill&D configuration will meet up to the WW^{++} level (2.6 m^2 PE^{-1}).

Based on this, it is clear that the area achieved by the CW is determined by the guidelines used. Therefore, the CW footprint reduction was also calculated using the first-order kinetic equation (assuming no background concentration, $C* = 0$ mg L^{-1}). The area obtained for each Duplex-CW configuration (using the final effluent concentration) was compared to that obtained for the VF CW alone (using the VF CW effluent concentration). In that regard, the Fill&D configuration saved almost no area during the 4 experimental periods for the treatment of COD, TSS, NH_4^+-N and TP. However, for TN, the Duplex-CW needed 2-3 times less space than that needed to reach the same effluent concentration using only a VF CW (Appendix C), similar to the area reduction for TN found by Zapater-Pereyra et al. (2014) when comparing a control HF CW with a hybrid HF CW (Chapter 4). Thus, the use of the Fill&D Duplex-CW served to reduce the systems area only for TN.

Commonly, VF CWs are generally sized in Europe with 1-3 m^2 PE^{-1} and HF CWs with 5 m^2 PE^{-1} (Vymazal, 2011) for the removal of organics and TSS, however that design is insufficient for nutrient removal (Babatunde et al., 2008). The Fill&D Duplex-CW area demand is included in the middle of this range. CWs like the French systems (e.g. Molle, 2014) and that described in Foladori et al. (2013) fit in the lower range (< 2 m^2 PE^{-1}). Probably a higher wastewater strength would have resulted in a smaller Duplex-CW area, as in Foladori et al. (2013) (using 74 g COD m^{-2} d^{-1}, double than in this study, reached the 1.5 m^2 PE^{-1}). For nutrients removal, areas of approximately 15-30 m^2 PE^{-1} and 40-70 m^2 PE^{-1} are suggested to be necessary to remove nitrogen (< 8 mg L^{-1}) and phosphorus (< 1.5 mg L^{-1}), respectively (Schierup et al., 1990 cited in Babatunde et al., 2008).

Based on this study findings (e.g. performance and space needed), the most appropriate Duplex-CW that can reduce the area requirements is the Fill&D configuration. This was possible due to the operational conditions (e.g. resting period and 1 d HRT in the VF CW) that enhanced aerobic removal processes. The observed solid accumulation confirmed the Fill&D configuration selection since the enhanced soil mineralization in the VF CW can minimize the risk of clogging of both compartments.

6.5. CONCLUSION

- The Fill&D Duplex-CW performed better than the StagB and FreeD systems due to the oxygen operational conditions and the HRT.
- The VF CW compartment contributed the most in the Duplex-CW overall treatment, since it was the first compartment in the Duplex-CW design and it showed high oxygen diffusion. The HFF contributed to further improve the VF CW treatment efficiency when needed.
- Artificial aeration improved effluent concentrations slightly, but not enough to show significant differences.
- Biofilm growth had a major impact in the HFF solids accumulation. The TSS load generated almost all the solids in the VF CW.
- The Fill&D Duplex-CW needed a 2-3 times lower area than what a single VF CW would have needed to reach similar TN effluent concentration. For other parameters (e.g.

COD, TSS and TP), the Duplex-CW did not contributed to the footprint reduction. The area requirement achieved was 2.6-3.4 m^2 PE^{-1}, lower than common European design (5 m^2 PE^{-1}) but still higher than many CWs.

■   The recommended Duplex-CW design should operate as the Fill&D configuration. Both compartments (VF CW and HFF) are relevant in the design. Aeration is not needed if treating up to the tested wastewater strengths.

CHAPTER 7.

MATERIAL SELECTION FOR A CONSTRUCTED WETROOF RECEIVING PRE-TREATED HIGH STRENGTH DOMESTIC WASTEWATER

A constructed wetroof (CWR) is defined in this study as the combination of a green roof and a constructed wetland: a shallow wastewater treatment system placed on the roof of a building. The foremost challenge of such CWR's, and the main aim of this investigation, is the selection of an appropriate matrix capable of assuring the required hydraulic retention time, the long-term stability and the roof load bearing capacity. Six substrata were subjected to water dynamics and destructive tests in two testing-tables. Among all the materials tested, the substratum configuration composed of sand, light expanded clay aggregates, biodegradable polylactic acid beads together with stabilization plates and a turf mat is capable of retaining the water for approximately 3.8 d and of providing stability (stabilization plates) and an immediate protection (turf mat) to the system. Based on those results, a full-scale CWR was constructed, which did not show any physical deterioration after one year of operation. Preliminary wastewater treatment results on the full-scale CWR suggest that it can highly remove main wastewater pollutants (e.g. COD, PO_4^{3-}-P and NH_4^+-N). The results of these tests and practical design considerations of the CWR are discussed in this paper.

This chapter is based on:
Zapater-Pereyra M., Dien van F., Bruggen van J.J.A., Lens P.N.L. (2013), "Material selection for a constructed wetroof receiving pre-treated high strength domestic wastewater", Water Science and Technology, 68 (10), 2264-2270.

7.1. INTRODUCTION

Cities are areas with a high and rapidly growing population density that, in most of the cases, generate a lot of environmental pollution. The little environmental awareness amongst the urban population coupled with their high economic capacity and high quality of life expectations enhance the city contamination (e.g. solid waste and wastewater production without appropriate treatment). Hence, adoption of localized wastewater treatment systems using green technologies is an important approach for sustainable development of cities. One major hindrance in urban areas is the land availability or, in some cases, the high costs of it. This warrants the need to find novel cost effective technologies that save space while promoting green growth. The possibility of using the roofs of buildings as a green space is a promising and eco-efficient strategy.

The most commonly employed use of roofs for solving different urban area problems has been Green Roofs (GRs) (Table 7.1). GRs are grass or sedum mats placed on the roofs of buildings having multi-layered arrangements (from bottom to top) - root barrier, drainage, filter, water retention, growing medium and vegetation. This arrangement can vary in material and thickness, depending on the type of GR and on the manufacturer (Bianchini and Hewage, 2012). A GR provides several urban environmental and operational cost-benefits as highlighted in Table 7.1. One potential benefit of GRs (not included in Table 7.1) is the treatment of wastewater. The GRs, with some modifications, can also serve as a wastewater treating biological filter; generating self-efficient buildings that are capable to recycle the water and thus, to reduce the water demand in cities. Therefore, it is believed that combining the properties of GRs and constructed wetlands (CWs) as a single system, i.e. a constructed wetroof (CWR), will provide significant advantages.

7.1.1. Roofs as wastewater treatment systems

To our knowledge, there are only a few studies that have addressed the challenge of treating wastewater on a roof: a natural wastewater treatment system at the John Deere plant in Mannheim (Germany) (Transfer, The Steinbeis Magazine, 2010), a Roof Garden study conducted by the Anhalt University of Applied Sciences (Bernburg, Germany; Thon et al., 2010) and a novel subsurface CW system called Green Roof Water Recycling System (GROW) patented in 2004 by Water Works, London, UK (Avery et al., 2006; Frazer-Williams et al., 2006; Winward et al., 2008). However, only the studies about the GROW system provide specific information about the design and performance.

A major limitation when constructing a natural wastewater treatment system on a roof is the load bearing capacity (LBC), the maximum weight a structure can resist. Thus, a CWR should be able to guarantee a proper wastewater treatment and at the same time to meet the weight restrictions to prevent structural damage. Therefore, this study aims to develop and to design a CWR for the treatment of pre-treated high strength domestic wastewater. The specific objectives of this research were the following: (i) to select the most suitable matrix for the CWR, (ii) to elucidate practical design considerations and, (iii) to present preliminary results of the efficiency of a CWR for domestic wastewater treatment.

Table 7.1 Problems presented in urban areas and the solutions provided by green roofs.

Urban area problems	Green Roof solutions
Scarce land/space availability.	Uses areas (roofs) commonly unused.
Lack of "green" spaces.	Provides green areas. Enhances the aesthetics of the area. Enhances biodiversity and aids relocating displaced animals.
Continuous rain creates flooding and collapses the sewage system.	Traps rain water shortly and releases it moderately to the sewage system. Provides area for evaporation of water. Decreases the amount of water to be treated in a wastewater treatment plant Regulates water runoff.
High (living) costs / Increased water and energy demand.	Reduces costs of air conditioning by covering the dark-roof with plants that trap solar energy. Reduces costs of repairing the roof due to the protection provided. Gives opportunity to store water, reducing water costs. Reduces costs of treating wastewater as less amount of (rain) water is discharged into the sewage system.
Air pollution (particulate and chemical).	Humidity provided by the plants traps air particles. Evaporation provides moist which has a higher binding capacity of particular matter. Leaves enzymes break down toxic chemicals into non-toxic components that can be used by the plant.
Poor microclimate and heat related illness/mortality.	Reduces heat island effect and related diseases.
Contributes to greenhouse gases emission.	Increases oxygen levels while reducing CO_2 quantities. Lowers CO_2 emission when reducing the conditioning system.
Population stress due to the rhythm, noise and concrete-architecture.	Green spaces provide calm and harmony among inhabitants. Green cities have positive effects on life quality. Lowers noise pollution by acting as insulation to the building.

7.2. MATERIALS AND METHODS

A CWR was built in April 2012 at the Van Helvoirt Groenprojecten (VHG) facilities (Tilburg, The Netherlands) (Appendix E). The company ECOFYT (Oirschot, The Netherlands) conducted the hydraulic design of the CWR as a subsurface horizontal flow CW. The CW volume obtained in the design calculations resulted in achieving a depth of 9 cm. The design features are displayed in Table 7.2. A key feature in the CWR design, and the focus of this study, was the selection of the CWR matrix that is capable of providing an appropriate hydraulic retention time (HRT) for the treatment of wastewater, while at the same time assuring long-term stability and meeting the roof LBC of 100 kg m^{-2} (this value does not include external factors such as rain, snow or people walking on the roof).

Table 7.2 Features of the constructed wetroof.

Parameter	Value
Constructed wetland type	Horizontal subsurface flow constructed wetland
Bed Area (m × m)	3.00 × 25.50
Depth (m)	0.09
Number of beds	4
Roof angle	14.3°
Flow type	Intermittent
Water pulses per bed per day	2
Volume of each pulse per meter of bed (L)	3.45-4
Wastewater type	Pre-treated domestic wastewater
Pre-treatment	Septic tank
(Expected) Retention time (d)	2-3
Resting period	1.5 d per week (Weekend). When roof temperature < 2°C.

7.2.1. Material selection: experimental set up and design

Six types of substrata (Table 7.3) were chronologically tested (November 2011 - March 2012) using two wooden testing-tables (length × width × depth: 3 × 1 × 0.2 m, slope 14.3°) placed indoors (Fig. 7.1). The materials used for the substrata were volcanic sand (0-4 mm), expanded polystyrene (EPS), crushed light expanded clay aggregates (LECA), fine sand (0-0.5 mm and 0-1 mm) and polylactic acid beads (PLA, biodegradable at temperatures above 70 °C).

Table 7.3 Constructed wetroof matrix options used in this study.

N°	Material	Material proportion (volume)	Presence of stabilization plate*	Presence of turf mat*	Matrix** depth (cm)	Matrix** dry weight (kg m^{-2})
1	Top layer (3 cm): Volcanic sand + Fine sand 0-1 mm	1:1	No	No	9	97
	Bottom layer (6 cm): Volcanic sand + EPS + Fine sand 0-1 mm	6:3:1				
2	9 cm of Volcanic sand + LECA	2:1	No	No	9	99
3	7.5 cm of Fine sand 0-0.5 mm + PLA	7:9	Yes	No, but included in matrix depth and weight	9	89
4	4.5 cm of Fine sand 0-0.5 mm	-	Yes	No, but included in matrix depth and weight	6	91
5	7.5 cm of Volcanic sand + LECA + PLA + Fine sand 0-0.5 mm	33:17:22:28	Yes	Yes***	9	90
6	4.5 cm of Volcanic sand + Fine sand 0-1 mm	1:1	Yes	No, but included in matrix depth and weight	6	77

*The stabilization plates (3.5 cm height, 3 kg m^{-2}) were embedded in the substratum and the turf mat (1.5 cm height, 15 kg m^{-2}) was placed on top of it.
**Matrix = substratum + stabilization plate + turf mat.
*** The turf mat was only added during the heavy-rain simulation experiment.

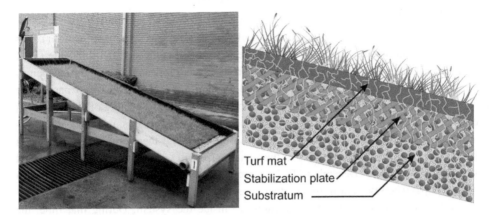

Figure 7.1 Testing-table (left) and the constructed wetroof selected matrix (right).

Sieve analyses were conducted to determine the particle size distribution curve of each substratum. Typically, this curve is calculated by the soil weight, however the weight of the materials used in this study varied from extremely light (e.g. PLA) to heavy (e.g. sand), hence, the particle size distribution curve was calculated by the material volume instead of material weight. From there on, the effective size (D_{10}), the uniformity coefficient and the hydraulic conductivity of each substratum were calculated. The hydraulic conductivity was calculated according to Hazen formula except when $D_{10} < 0.1$ mm. In that case, Kenney's formula was used (Hill and Benson, 1999). Additionally, the volumetric dry mass and the porosity were measured.

The substrata N° 3-6 (Table 7.3) included the presence of "stabilization plates" within the media and considered the use of a "turf mat" (sod of grass) on top (1.5 cm height). From this point forward, the word "matrix" in this study refers to the combination of substratum, stabilization plate and turf mat (Fig. 7.1). Two main considerations had to be taken into account: (i) the total matrix weight of the substratum and wastewater on the roof should be $\leq$ 100 kg m^{-2} and (ii) the total matrix depth should be (at least) 9 cm depth (after compaction) (Table 7.2).

The water hydrodynamics were tested in each substratum by adding tap water (3.45-4 L) twice a day to the testing-tables. The water volume, in and out, was recorded twice a day for approximately 8-10 d. Destructive tests such as weight application (by walking on the compacted substratum) and heavy-rain simulation (only in Substratum N° 5) were conducted to test the physical stability of the matrix. For Substratum N° 5, a tracer test was conducted during 8 d with NaCl (35 g L^{-1}) to estimate the HRT. Electric conductivity (EC) was recorded (with a data logger) every 15 min.

7.2.2. **Wastewater in the full-scale constructed wetroof**

Approximately 120 L d^{-1} of pre-treated (septic tank) high strength domestic wastewater, coming from the office building (i.e. kitchen sink, toilet, hand wash basin and shower), was pumped daily (intermittent mode) to each of the 4 beds that composed the full-scale CWR (Table 7.2). Effluent samples of the 4 beds were taken as replicates. Sampling campaigns were conducted during July 2012 (3 times) and October 2012 (5 times) for all parameters except for total nitrogen (TN) and total phosphorus (TP). In 2013, sampling of all parameters was conducted once in April, and twice in May and June. In many cases, the effluent buckets of each bed were found empty (due to evapotranspiration) and therefore no water analyses could be done, resulting in variations in the sample size. No water sampling was conducted during the winter period (December 2012 - March 2013). The CWR was designed to switch off (the pumps) when the substratum temperature, that was permanently recorded, was < 2°C in order to avoid the freezing of the wastewater inside the system. During that time, the wastewater was discharged to the sewer system.

Water samples at the influent and effluent of each bed were analyzed for pH, EC and dissolved oxygen (DO) by the electrometric method; COD by the potassium dichromate oxidation method; TSS by the gravimetric method and PO$_4^{3-}$-P and TP by the vanadomolybdate method following standard procedures (APHA, 2005). NH$_4^+$-N, by the dichloroisocyanurate method, following the Dutch Standards NEN 6472 (1983) and NO$_3^-$-N was analyzed by ion chromatography. TN was measured by the persulfate method (Koroleff, 1983).

7.3. RESULTS AND DISCUSSION

7.3.1. **Substrata characteristics and testing-table tests**

Table 7.4 summarizes the soil results of each substratum considered in this study. They showed high variations, among each other, with respect to size range, volumetric dry mass, hydraulic conductivity, D$_{10}$ and uniformity coefficient. However, all had similar porosity.

The measurements of water volume in and out (data not shown) in Substratum N° 1 and 2 showed that the water was, in both substrata, leaving the system very fast (approximately 0.34 and 0.25 L h^{-1} for Substratum N° 1 and 2, respectively), indicating that the HRT was less than a day. These results are supported by the high levels of hydraulic conductivity (131-175 m d^{-1}) (Table 7.4). The reason behind the use of Substratum N° 1 and 2 was to assure a very light material (EPS and LECA, respectively) that could reach the 9 cm depth. Nevertheless, these substrata were too coarse and the water was going out of the system almost immediately. Such a short HRT in a shallow filter bed is not enough for an appropriate wastewater treatment.

Table 7.4 Main soil parameters of the substrata used in this study.

Parameter	Substratum N°					
	1 (top/bottom)	2	3	4	5	6
Size range (mm)	0.063-2.0/0.063-8.0	0.02-8.0	0.02-4.0	0.02-1.0	0.02-8.0	0.063-4.0
D_{10} (mm)	0.42/0.41	0.45	0.09	0.07	0.13	0.27
Uniformity coefficient	3.11/4.51	3.67	11.05	3.00	10.94	3.17
Hydraulic conductivity (m d^{-1})	153.9/145.2	175.0	0.46	0.3	14.2	60.7
Porosity (%)	44.3/42.3	44.9	41.3	38.5	45.2	45.4
Volumetric dry mass (kg m^{-3})	1255/982	1096	952	1616	954	1317

Substratum N° 1 - Top layer: Volcanic sand + Fine sand 0-1 mm, Bottom layer: Volcanic sand + EPS + Fine sand 0-1 mm; Substratum N° 2 - Volcanic sand + LECA; Substratum N° 3 - Fine sand 0-0.5 mm + PLA; Substratum N° 4 - Fine sand 0-0.5 mm; Substratum N° 5 - Volcanic sand + LECA + PLA + Fine sand 0-0.5 mm and Substratum N° 6 - Volcanic sand + Fine sand 0-1 mm.

The destructive tests (weight application) revealed that the bare substrata, even well compacted, were not able to provide the stability required to assure a long lifespan of the system. Therefore, the use of a recycled-plastic stabilization plate embedded in the substratum and a turf mat on top of it were considered (Fig. 7.1). The benefit of using stabilization plates close to the surface, following FLL guidelines (2002), was immediately noticed after repeating the destructive test (no deformations were visible on the substratum surface). The turf mat major contribution is explained later in this section.

Another substratum trial (N° 3 and 4) was conducted with the purpose of increasing the HRT and the stability of the system (by including stabilization plates). However, after a few days of water application, short circuits were noticed in both systems. The hydraulic conductivity of such substrata was too small (Table 7.4) that it was difficult for the water to penetrate through the substratum and instead it searched for easy paths inducing preferential flows. To confirm this behavior, the test was repeated with both substrata and again, after a few days, similar preferential flow channels developed.

A final trial (Substrata N° 5 and 6) was done with a coarser substratum that still guarantees the necessary HRT but at the same time allows the water to flow through it without causing damage. During water addition, both substrata behaved similarly: no visual sign of preferential flow or short circuiting. As Substratum N° 6, due to weight requirements, had to have a 4.5 cm depth, it was not the preferred substratum. Substratum N° 5 instead had the desired specific weight so the necessary depth (9 cm, Table 7.2) was met. The combination of all the different materials - with different sizes - created a more stable substratum (Table 7.3). The smaller particles fit the pores of the coarser material, while at the same time the hydraulic conductivity (Table 7.4) of this substratum appeared to guarantee the desired flow of water without the formation of preferential flows.

It was concluded that Substratum N° 5 was the best tested option and therefore a tracer test was conducted to quantify the exact HRT. The results of the tracer test indicated that the HRT in the testing-table was approximately 3.8 d.

The final step of the material selection process was a heavy-rain simulation test conducted for Substratum N° 5. It was hypothesized that the bare substrata were not able to resist precipitation, wind or snow until the plants would have developed. Therefore, before the rain test, an immediate vegetation cover on top of the system - a turf mat - was placed (Fig. 7.1). The heavy-rain simulation test confirmed that, without the presence of the turf mat, the substratum would have immediately failed. However, the turf mat does not guarantee complete success of the system, but protects it and provides an immediate presence of plants for further roots colonization.

The total final matrix selected for the construction of the CWR was: a substratum of volcanic sand + LECA + PLA + fine sand, the stabilization plate and the turf mat.

7.3.2. Wastewater treatment

Table 7.5 shows the wastewater parameters concentrations in the influent and effluent of the full-scale CWR. The system shows removal efficiencies exceeding 86% for TSS, 85% for COD, 99 % for NH_4^+-N and 68% for PO_4^{3-}-P, similar to the performance of conventional horizontal flow CWs (Vymazal 2005; García et al., 2010) and to that of the GROW system (for TSS and COD) when treating low strength wastewater (Winward et al., 2008). The DO of the effluent increased to aerobic conditions (3 mg L^{-1}); suggesting, together with the high NH_4^+-N removal, that nitrification is taking place in the shallow bed. Furthermore, TN is highly removed (97%) opposite to what usually occurs in conventional CWs (from 40-55%, García et al., 2010), suggesting a high denitrification potential of the full-scale CWR.

Table 7.5 Water quality in the influent and effluent of the full-scale constructed wetroof.

Parameter	Influent	n	Effluent	n	% removal
pH	7.4-8.7	6	7.4-8.4	5	-
Dissolved oxygen (mg L^{-1})	0.3 ± 0.1	6	3.2 ± 1.8	5	-
Electric conductivity (μS cm^{-1})	2162 ± 161	6	1438 ± 344	5	-
Total suspended solids (mg L^{-1})	186 ± 22	11	26 ± 5	8	86
Chemical oxygen demand (mg L^{-1})	859 ± 76	11	129 ± 31	9	85
NH_4^+-N (mg L^{-1})	187 ± 9	12	0.2 ± 0.1	9	99.9
NO_3^--N (mg L^{-1})	0.08 ± 0.02	4	8.05 ± 3.28	7	-
Total nitrogen (mg L^{-1})	225 ± 8	4	7 ± 2	2	97
Total phosphorus (mg L^{-1})	27 ± 3	5	8.7 ± 2.7	6	68

Results are mean ± standard error, except for pH.

7.3.3. **Practical design considerations**

Choosing an appropriate substratum for the full-scale CWR is the most important task for its long-term success. During this study some key points were tried to achieve and several lessons were learnt:

▪ Limited LBC of the building/roof: a parameter that limits the amount and type of material to be used. LECA, EPS and PLA beads were key materials to decrease the matrix weight.

▪ Enough surface area for biofilm development and appropriate water purification: in a very shallow bed (9 cm in this study), very fine material should be used to provide enough surface area. Fine material usually are heavy (e.g. sand) and can compromise the LBC of the building. Light and high-surface-area materials (e.g. LECA and PLA beads) are key materials for this purpose.

▪ Wastewater treatment: a certain substratum volume with an appropriate HRT is necessary for a suitable treatment. When weight restrictions exist (limited LBC), it is important to mix common (e.g. sand) and light materials with a particle size range that assures a proper hydraulic conductivity (without preferential flow), and therefore a long enough HRT.

▪ Increased stability: the system, especially on sloped-roofs, requires to be stable to support seasonal events (e.g. rain) and walking of people (during maintenance). The material alone could not guarantee such stability and therefore it is of importance to include stabilization plates.

▪ System sustainability: the use of natural materials is desired but without man-made light weight beads such as LECA, EPS and PLA, weight requirements cannot not be met. Therefore, the biodegradable PLA beads were preferred over EPS.

▪ CWR construction: it is desired that a CWR construction occurs during a period without excessive precipitation, preferably until the roots have developed and stabilized the soil matrix. The use of a turf mat is a crucial feature to minimize environmental (e.g. rain and snow) impacts.

The CWR is a promising wastewater treatment system designed for spaces commonly unused, namely roofs. However, the current design (9 cm depth) requires approximately 76 m^2 PE^{-1} (population equivalent) (1PE = 60 g BOD_5, Eq. 2.1), a relatively high value for a single household application. Therefore, such a design is mainly recommended for large buildings with enough space and low PE (e.g. shopping malls or office buildings). For single households or housing complexes two approaches are recommended: (i) a more stable structural design (prior construction) of the house to increase the LBC and therefore to allow a deeper CWR matrix than that described in this study and (ii) a new type of urban planning where single households nearby large buildings with CWRs deliver their wastewater for treatment and subsequent recycling of water in common green areas.

7.4. CONCLUSIONS

▪ The configuration composed of material with different particle sizes (volcanic sand + LECA + PLA beads + fine sand), together with the stabilization plates and turf mat is capable of retaining the water for an appropriate amount of time (approximately 3.8 d).

▪   Preliminary wastewater treatment results conducted in the full-scale CWR indicate that the system - with the chosen matrix - has a high potential for wastewater purification. Currently, the full-scale CWR is successfully working over one year without showing any physical deterioration of the matrix.

CHAPTER 8.

CONSTRUCTED WETROOFS: A NOVEL APPROACH FOR THE TREATMENT AND REUSE OF DOMESTIC WASTEWATER AT HOUSEHOLD LEVEL

The lack of space in urban areas can be a reason for their lack of green areas and sanitation provisions. Since roofs represent common unused space, they become an option to locate natural treatment systems such as constructed wetlands (CW). A conventional CW is too heavy to be placed on a roof; hence its design must be changed without altering its treatment properties. Thus, the concept of a constructed wetroof (CWR) was developed, a 9-cm-depth system composed of sand, organic soil, light expanded clay aggregates (LECA), polylactic acid beads (PLA) and roots. Due to the shallow depth and unsheltered location, the CWR is susceptible to changes in its physical and hydrological properties (such as a hot day causing severe evapotranspiration that dries out the filter media and the plants) that may lead to treatment failures. Therefore, this study investigates the capacity of a CWR to treat domestic wastewater from an office building in The Netherlands. The results showed that the CWR is a highly aerobic system that treats domestic wastewater beyond the levels required by local discharge quality guidelines for organic matter, solids, NH_4^+-N, total nitrogen and total phosphorus (> 79% for all parameters tested). The roots, organic soil and sand provided the majority of the treatment; while the LECA and PLA were used mainly to provide volume without increasing significantly the weight. LECA also played a significant role in phosphorus removal. The operation of the CWR was highly weather dependent, however the treatment efficiency was not affected by it. Moreover, the rain had a minor effect on washing away the nutrients retained in the CWR media. The CWR provided a resilient and efficient domestic wastewater treatment that requires, as it is placed on a roof, 0 m^2 per population equivalent.

This chapter is based on:
Zapater-Pereyra M., Lavrnić S., Dien van F., Bruggen van J.J.A., Lens P.N.L., "Constructed wetroofs: a novel approach for the treatment and reuse of domestic wastewater at household level", Submitted to Journal of Environmental Management.

8.1. INTRODUCTION

A green roof (GR) is a vegetated surface located on the roof of buildings that contributes to solving some city problems such as the lack of green areas, the heat island effect and air pollution (Zapater-Pereyra et al., 2013; Berardi et al., 2014) since roofs represent approximately 32% of the horizontal built-up surface in urban areas (Oberndorfer et al., 2007). The use of vegetation on roofs exists since 500 B.C. (e.g. the hanging gardens of Babylon), but it was only in the 20th century when modern GRs started to be used in Germany (Getter and Rowe, 2006; Oberndorfer et al., 2007). The concept was since then widely expanded for research and application.

It is repeatedly documented that population growth results in increasing urbanization (where concrete architecture predominates over green open spaces), depletion of fresh water sources, production of wastewater and thus a need for more sanitation. Hence driving the mindset towards the use of GRs for wastewater treatment is a valid (and needed) argument to solve the consequences of urbanization. Research has mainly focused on understanding the types of GRs, temperature-related services, substratum and plant selection, exposure to wet/dry conditions, water runoff, storm water quality, air pollution, roof longevity, urban heat island, noise reduction, landscape properties, costs (e.g. construction, energy-related), policies and ecosystem services. However, only a few studies have shown the potential of GRs for wastewater treatment (Avery et al., 2006; Frazer-Williams et al., 2006; Winward et al., 2008; Thon et al., 2010; Transfer, The Steinbeis Magazine, 2010; Zapater-Pereyra et al., 2013). Such type of systems from now on will be referred to as constructed wetroofs (CWR), as they combine the properties of GRs and constructed wetlands (CW).

A CWR must meet the load bearing capacity (LBC) of the building (Section 7.1.1) and provide an appropriate hydraulic retention time (HRT) for the treatment of the wastewater. This was investigated in a previous study (Zapater-Pereyra et al., 2013 - Chapter 7). Due to the shallow depth and unsheltered location, vegetated roofs tend to be highly susceptible to deterioration due to environmental conditions (e.g dry, rainy, cold or windy period) that could compromise their stability (thus, lifespan) and, if they were designed for wastewater treatment (i.e. CWR), it could compromise the treatment performance. For instance, bacterial communities and the water path along the system (thus, the HRT) can be affected by environmental fluctuations.

Up to date, no study has thus far addressed the hydrology issues of a CWR and direct linked it with its treatment performance. Therefore, this study was divided in three parts: (i) water quality, (ii) nutrient accumulation in and microbial activity of the media and (iii) water flow path during different environmental conditions. It aimed at understanding the potential of CWRs for wastewater treatment while being exposed to different hydrological conditions.

8.2. MATERIALS AND METHODS

8.2.1. Constructed wetroof description

The CWR was built in April 2012 on the roof of an office building of the company Van Helvoirt Groenprojecten in Tilburg (The Netherlands) (Zapater-Pereyra et al., 2013) (Fig. 8.1A, Appendix E). It was designed as a shallow 9 cm deep horizontal flow (HF) CW by the company ECOFYT (Oirschot, The Netherlands) and was divided in four identical beds (3.0 × 25.5 m) with a slope of 14.3° (Fig. 8.1A) and a HRT of ~3.8 d (Section 7.3.1). The substratum used was a mixture of two types of sand, light expanded clay aggregates (LECA) and polylactic acid (PLA) beads. Stabilization plates (3.5 cm height) were embedded at the top of the 7.5-cm substratum and turf mats (or sod of grass, 1.5 cm height) were placed on top (Fig. 7.1, Chapter 7). The turf mat (Kuypers, The Netherlands) was grown in sandy soils that were highly fertilized (such soil from now on will be called "organic soil") and the composition of the grass seeds was 20% *Lolium perenne*, 50% *Festuca rubra* and 30% *Poa pratensis* according to the manufacturer's specifications (kuypers-graszoden.nl). More information of the CWR design can be found in Zapater-Pereyra et al. (2013) (Chapter 7). From this point onwards, the terms "turf mat", "substratum" and "media" will refer to, respectively, organic soil + grass roots, sand + LECA + PLA beads and turf mat + substratum + grass roots.

The domestic wastewater from the building was collected in a septic tank, from where it overflowed into the inlet tank (pump sump) of the CWR. Four pumps (one for each bed), submerged in the inlet tank, were controlled by a switchboard that sequentially activated each pump (one pump cycle ~93 L) depending on the wastewater production. The wastewater was pumped to the beds via pressure pipes. The influent pipe, with holes (Ø ~6 mm) every 1 m, released the wastewater to a channel, from where it could infiltrate in the CWR bed. The effluent pipe had many more random holes (Ø ~5.5 mm) that collected the water from the bed and conveyed it to an effluent tank (Mixed Effluent) from where it was reused for toilet flushing. Surpluses were directed (by gravity) to an infiltration pond. For research purposes, the effluent was also collected in separate buckets in order to have water from each bed (Bed Effluent).

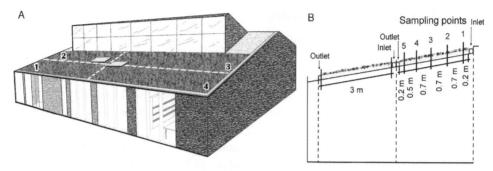

Figure 8.1 Schematic representation of the office building with the constructed wetroof (A) and a section of it indicating the five sampling points along a bed length (B). The dashed white lines in graph A are displayed only for an easy visualization of the four identical beds (indicated by the numbers).

8.2.2. Experiments and sampling

Hydrology. To understand the water path along the bed length, electrodes were inserted in the CWR in 5 points along the bed length to record the moisture content. Point 1 was located at 0.2 m from the inlet and the others (Point 2-5) at 0.7, 0.7, 0.7 and 0.5m from the preceding point (Fig. 8.1B). They were further connected to a data logger (Em5b Data Collection System, Decagon Devices, USA) that indirectly measured the dielectric permittivity of the media (every 15 min) and provided an output in mV. As water greatly influences the dielectric permittivity, the mV output will increase when the sample is wetter. The 5 electrodes were rotated from bed to bed (from April until December 2013, n = 23284). The average data is presented and few dates are plotted as well to visualize the water path on a dry (sunny) and a wet (rainy) day. The moisture content in the 5 points was compared statistically (with SigmaPlot 12.3 software) using the Non-parametric Kruskal-Wallis One Way ANOVA on Ranks (as the data did not meet normality).

To calculate the water balance, the volume of the water entering the system was monitored in working days (July 2012 - January 2014, a total of 551 days). During severe hot days, the CWR experienced total dryness, hence the decision of using randomly sprinklers for a short period in summer 2013. The water applied by the sprinklers was also taken into account for the water balance. The water exiting the system was divided into the water for toilet reuse and the overflow to the infiltration pond. The water meters for both were installed later; therefore the recording time period was from June 2013 to January 2014. Rain data was taken from the Royal Netherlands Meteorological Institute website (www.knmi.nl, from the nearby Tilburg weather station at approximately 1.5 km away).

Water. Sampling was conducted, for influent and effluents, 3 times in April 2014 and 4 times in May 2014. Those results were joined with previous water quality data reported in Zapater-Pereyra et al. (2013) (Chapter 7) (sampled in July and October 2012 and April-June 2013), except for 5-d biological oxygen demand (BOD₅) that was only done in this study. All the raw data was averaged and the results are presented in Table 8.2. The CWR did not operate when bed temperatures were below 2°C (winter time, approximately mid December to mid March), since it was hypothesized that the shallow bed could not prevent the freezing of the wastewater inside the system. During that time, the wastewater was discharged to the sewer system.

CWR media. The media was sampled at the 5 points along the length of each bed (Fig. 8.1B) in order to analyze potential nitrification and denitrification rate (PNR and PDR, respectively; 5th May 2014, n = 4) and microbial activity (2 beds 14th May 2013 and all beds 22nd May 2013, n = 6). Microbial activity was determined for the turf mat and the substratum separately. This separation was done with the intention of homogenizing the results after realizing that they were extremely different (factor of 10), even between the replicates, when both layers were mixed. The nutrient content was also sampled at the 5 points along the length on the 23rd April and 14th May 2013 but only the average data for the whole CWR is presented (n = 40).

Three 10 × 10 cm bed samples were taken from the CWR on the 23rd April 2014. From each, a subsample was used to measure the "present" (after 2 years of operation) microbial activity of the substratum, the turf mat and the single materials (i.e. sand, LECA, PLA and roots) (n = 3). One of the 10 x 10 cm bed samples was dried and the materials were separated (i.e. sand, LECA, PLA, root and turf mat). The weight and volume of each material was quantified and their percentage from the total was calculated. These results, combined with the microbial activity, were used to calculate the CWR total activity and thus to understand which material provided the majority of the domestic wastewater treatment in the CWR. A specimen of the "original" mixture (before using it in the CWR) was also separated per material and used to analyze microbial activity (n = 1), except for the roots and turf mat, to discard the idea that the present activity was due to the original materials.

The effect of rain on the nutrients adsorbed in the media was tested by measuring the total nitrogen (TN) and total phosphorus (TP) content before and after a rain event. Three random sampling points were chosen per bed and care was taken to sample both, before and after the rain, next to each other. Three trials were conducted in random beds, each with a different rain intensity: 4.4 mm d^{-1} (real rain, n = 3), 2.3 mm d^{-1} (real rain, n = 12) and 18.1 mm d^{-1} (rain simulation with sprinklers, n = 12), in September, October and December 2013, respectively. Rain data was taken as well from the weather station Tilburg for the real events and with a pluviometer for the simulated rain. The rain intensities were called trace (0.1-1 mm d^{-1}), light (1-4 mm d^{-1}), light-moderate (4-16 mm d^{-1}) and moderate-heavy (16-32 mm d^{-1}) following the categories in Homar et al. (2010). To compare the nutrient concentration before and after the rain event, a paired T-test (2-tailed, p<0.05) was conducted, using SigmaPlot 12.3 software. If normality assumption was not met, the Wilcoxon Signed Rank Test was used.

8.2.3. **Analytical methods**

Water. Influent and effluent water was tested for several parameters following APHA (2005), unless otherwise stated. The pH, temperature, dissolved oxygen (DO) and electrical conductivity (EC) were measured using the electrometric method. Ammonium nitrogen (NH_4^+-N) was analysed by dichloroisocyanurate method (NEN 6472, 1983). Nitrate nitrogen (NO_3^--N) was analysed using an ion chromatograph (ICS 1000, DIONEX, USA). TN was measured by the persulfate method, while TP sample preparation and digestion procedure were similar to TN, but samples were analysed with the vanadomolybdophosphoric acid colorimetric method. Chemical oxygen demand (COD) analysis was done using an open reflux titrimetric method, BOD_5 by the oxygen electrode method and total suspended solids (TSS) by the gravimetric method.

CWR media. Nutrient content (TN and TP) in the dry media (70 °C) was analyzed by digestion with H_2SO_4/Se/salicylic acid and H_2O_2 (Walinga et al, 1989). Digestion was followed by NH_4^+-N and PO_4^{3-}-P analysis by the dichloroisocyanurate method (NEN 6472, 1983) and the ascorbic acid spectrophotometric method (APHA, 2005), respectively. Microbial activity by the fluorescein diacetate (FDA) assay followed the protocol described in Adam and Duncan (2001).

The PNR and PDR of the media were determined as described by Xu et al. (2013) with some modifications. Briefly, for both tests, samples (~15 g) were put in 250 mL glass bottles containing 120 mL of the respective incubation solution following the recipe of Xu et al. (2013). The incubation solution had an initial concentration of ~300 mg NH_4^+-N L^{-1} in the nitrification test and ~250 mg NO_3^--N L^{-1} in the denitrification test. In the nitrification test, the bottles were then placed on a rotary shaker at 175 rpm and 30°C; while for the denitrification test the bottles were flushed with nitrogen gas (to create anaerobic conditions), capped and incubated at 30°C. Sampling for both methods was conducted at time 0 and 48 h and the nitrate concentration was analysed. Results were calculated according to Xu et al. (2013). The PDR results were then multiplied by 85% as a correction factor for some nitrate-ammonifiers present in the media converting NO_3^--N to NH_4^+-N (unpublished results).

8.3. RESULTS

8.3.1. Water path along the bed length and water balance

The average moisture content results along the bed length were (mean ± Std. error; in mV): 424 ±0.5 for Point 1, 388 ± 0.5 for Point 2, 412 ± 0.6 for Point 3, 419 ± 0.6 for Point 4 and 396 ± 0.4 for Point 5. All were statistically different to each other ($p<0.05$).

A close observation at the collected data revealed that during dry days, the inlet (Point 1) was the most wet area in the bed. The other points showed similar values among them with random minor variations in their degree of moisture content. Figure 8.2A shows a typical example of the bed moisture content during dry days. During wet days, the CWR hydrology varied depending on the rain category. A trace and light rain had no visible effect on the bed moisture content (Fig. 8.2B), a light-moderate rain created an immediate moisture content increment without changing the order of the points (Fig. 8.2B-C) and a moderate-heavy rain immediately increased the bed moisture content of all points to the same extent (Fig. 8.2B). It should be noted that Figure 8.2 only represents a short example among the total collected data.

The water balance (Table 8.1) showed that the total input of water is about 1212 L d^{-1}, while only 458 L d^{-1} was measured as output (for reuse and overflow). The difference (754 L d^{-1}) was considered to be evapotranspiration and losses (e.g. pipe leakages and spills of wastewater due to a pump malfunction, pumping more than necessary). From the input, rain accounted for a 55% of the total input while the domestic wastewater was slightly less (41%). From the output, 143 L d^{-1} was reused for toilet flushing and twice that amount (315 L d^{-1}) was discharged in the infiltration pond.

Table 8.1 Water balance in the constructed wetroof (mean ± Std. error) obtained in a period of 551 days.

	Flow (L d^{-1})	% of Total Input	n
Input			
Domestic wastewater	499 ± 65	41.2	232
Rain	668 ± 71	55.1	235
Sprinkler water*	45	3.7	-
Total Input	*1212*	-	-
Output			
Reuse (toilet flushing)	143 ± 6	11.8	89
Overflow (to infiltration pond)	315 ± 94	26.0	89
Subtotal Output	*458*	*37.8*	-
Evapotranspiration & losses**	754	62.2	-

* The sprinkler water flow was calculated by summing the total amount of water added with the sprinklers in the study (25 m^3) and dividing it by the 551 days.
**Calculated value: (Total Input) - (Subtotal Output).

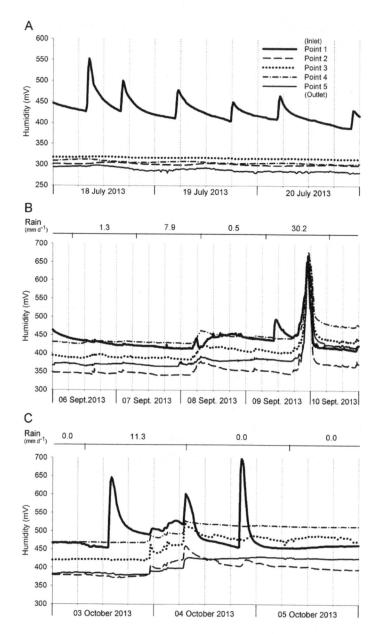

Figure 8.2 Humidity fluctuations along the length of the bed in dry (sunny) (A) and wet (rainy) periods (B and C). Daily rain data is collected from 08:00 to 08:00 at Tilburg weather station.

8.3.2. Wastewater treatment

Table 8.2 shows the wastewater quality in the influent and effluent of the CWR. The pH of the water did not change with the treatment while the oxygen concentration increased from 0.2 to approximately 5-6 mg L^{-1}. All major wastewater quality parameters were highly removed (79-99.8%) with effluent concentrations lower than: 37 mg L^{-1} for TSS, 155 mg L^{-1} for COD, 11 mg L^{-1} for BOD$_5$, 0.7 mg L^{-1} for NH$_4^+$-N, 32 mg L^{-1} for TN and 4 mg L^{-1} for TP. NO$_3^-$-N increased from negligible values to approximately 14-17 mg L^{-1}. Values from the mixed and bed effluent varied to different extents, but were in general in the same order of magnitude.

Table 8.2 Water quality (Mean ± Std. error, except for pH) and removal efficiency (%Rem.) in the influent and effluent of the constructed wetroof. Adapted and updated from Zapater-Pereyra et al. (2013) (Table 7.5). Units are in mg L^{-1} unless stated otherwise.

Parameter	Influent	n	Mixed Effluent	n	% Rem.	Bed Effluent	n	% Rem.
pH (unitless)	7.4-8.7	14	7.2-7.8	14	-	6.6-8.5	17	-
Dissolved oxygen	0.2 ± 0.1	14	5.2 ± 0.8	14	-	6.0 ± 0.8	17	-
Electric conductivity (μS cm^{-1})	2660 ± 138	14	564 ± 21	14	-	888 ± 142	17	-
Total suspended solids	190 ± 15	18	17 ± 2	14	91.3	37 ± 9	19	80.7
Chemical oxygen demand	754 ± 50	20	132 ± 16	16	82.5	155 ± 17	21	79.4
Biological oxygen demand	217 ± 16	6	7 ± 1	6	96.6	11 ± 1	10	94.9
NH$_4^+$-N	205 ± 8	19	0.7 ± 0.1	16	99.7	0.4 ± 0.1	21	99.8
NO$_3^-$-N	0.2 ± 0.1	14	13.9 ±	11	-	16.7 ± 3.3	17	-
Total nitrogen	251 ± 8	12	19 ± 2	12	92.6	32 ± 6	14	87.1
Total phosphorus	30 ± 1	13	1 ± 0	13	97.2	4 ± 1	18	86.3

8.3.3. Activity of the media

Figure 8.3 shows that the microbial activity in the turf mat is higher than that in the substratum. Moreover, a slight decrease in activity occurred from inlet (Point 1) towards the outlet (Point 5). The microbial activity measured in the "present" materials (after 2 years of operation) showed that a high activity was concentrated in the sand (47 μg$_{fluorescein}$ mL^{-1}$_{material}$ h^{-1}), while much lower activity was encountered in the LECA and PLA (6 and 2 μg$_{fluorescein}$ mL^{-1}$_{material}$ h^{-1}, respectively) (Table 8.3). However, in the roots and the turf mat (organic soil + roots), the microbial activity is by far higher than in the other materials, over 3 times that of sand. Microbial activity values of the original samples were negligible (< 2 μg$_{fluorescein}$ mL^{-1}$_{material}$ h^{-1}) (Table 8.3).

Table 8.3 Microbial activity (Mean ± Std. error) of the different constructed wetroof (CWR) materials from the original (n = 1) and present (n = 3) sample, the weight and volume that they occupy in the CWR and their total activity in the CWR.

	Microbial activity ($\mu g_{fluorescein}$ mL$^{-1}_{material}$ h^{-1})		Original volume occupied in the CWR (based on design values)	Present weight & volume occupied in the CWR (from 10 × 10 cm sample)		CWR total activity*** ($g_{fluorescein}$ h^{-1})
	Original (1)	Present (2)	% of mL (3)	% of g (4)	% of mL (5)	Present (6)
Substratum	1	40 ± 10	-	-	-	-
Sand*	2	47 ± 16	50.8	73.5	45.2	658
LECA	1	6 ± 1	14.2	7.9	14.9	28
PLA beads	0	2 ± 0	18.3	2.8	15.8	10
Roots	ND	150 ± 13	2.5	1.5	11.6	539
Organic soil	ND	216 ± 70**	14.2	14.3	12.5	875

ND, not determined.

* The large particles of LECA were completely removed from the sand, but some small (crushed) particles remained.

** The value refers to the whole turf mat (organic soil + roots) microbial activity.

*** Calculation: (Column 2) × (CWR volume) × (Column 5); except for the Organic soil value which was subtracted from the turf mat microbial activity (**) minus the root activity in the turf mat only. The total CWR volume was 30957000 mL.

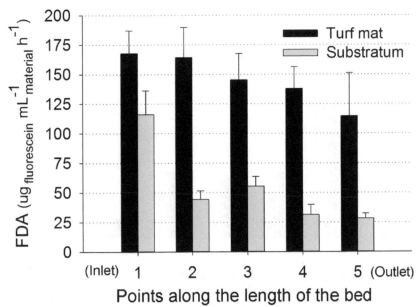

Figure 8.3 Microbial activity, using the fluorescein diacetate assay (FDA), in the turf mat and the substratum along the length of the bed. *Note: to express FDA per gram of material, values displayed in the figure should be divided by 1.571 g mL^{-1} for the substratum and 0.510 g mL^{-1} for the turf mat.*

The PNR showed the highest value at the inlet (0.37 mg N L$^{-1}_{media}$ h^{-1}) and the lowest at the outlet (0.05 mg N L$^{-1}_{media}$ h^{-1}) (Fig. 8.4). Middle points were similar, in the order of 0.16 mg N L$^{-1}_{media}$ h^{-1}. The PDR showed similar values for all points (~33 mg N L$^{-1}_{media}$ h^{-1}), except at the inlet (Point 1) that was slightly lower (29 mg N L$^{-1}_{media}$ h^{-1}) (Fig. 8.4). The weighted-mean PNR and PDR for the whole CWR were 0.2 and 34.8 mg N L$^{-1}_{media}$ h^{-1}, respectively. The "actual" nitrification and denitrification rates for the whole system were also calculated (0.04 mg N L$^{-1}_{media}$ h^{-1}, for both) and 1-3 orders of magnitude lower than the potential values (Fig. 8.4).

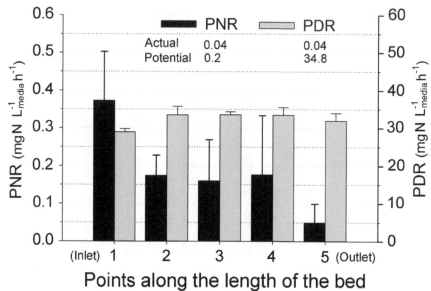

Figure 8.4 Potential nitrification rate (PNR) and potential denitrification rate (PDR) in the media along the length of the bed and the mean Actual (calculated from the effluent quality results) and the Potential (calculated from PNR and PDR) rates.

8.3.4. Nutrient content and balance

On average, the media contained 0.29 mg N g$^{-1}_{dry\ media}$ and 0.33 mg P g$^{-1}_{dry\ media}$. Those values were used to calculate the nutrient balance in the CWR (Table 8.4). For nitrogen, out of the 45.8 kg yr^{-1} at the influent, only 3.1 kg yr^{-1} (6.8%) and 3.8 kg yr^{-1} (8.3%) were found in the effluent mixed and media, respectively. For phosphorus, 0.1 kg yr^{-1} (2.6%) and 2.8 kg yr^{-1} (51.4%) from the influent (5.5 kg yr^{-1}), were found in the effluent mixed and the media, respectively.

The nitrogen and phosphorus content in the media were measured before and after different rain events to determine if a rain washes out immediately the nutrient retained (Table 8.5). Results showed that there were no significant differences, neither in nitrogen nor in phosphorus content, before and after the rain event, independently of the rain intensity (p>0.05, Table 8.5).

Table 8.4 Nutrient balance in the constructed wetroof.

	Nitrogen (kg yr^{-1})	% of influent Nitrogen	Phosphorus (kg yr^{-1})	% of influent Phosphorus
Influent	45.8	-	5.5	-
Effluent mixed	3.1	6.8	0.1	2.6
Media*	3.8	8.3	2.8	51.4
Losses	*38.9*	*84.9*	*2.5*	*46.0*

*The nutrients found in the original substratum where subtracted from the nutrient values in the media displayed in this table. Turf mat original nutrient content was not determined, thus it was not subtracted.

Table 8.5 Nutrient concentration fluctuations in the constructed wetroof media before and after rain events.

Rain intensity (mm d^{-1})	Rain category	n	Wastewater cycles added during experiment*	Mean (± Std. error) nitrogen (mg N g^{-1} $_{media}$)		Mean (± Std. error) phosphorus (mg P g^{-1} $_{media}$)	
				Before rain	After rain	Before rain	After rain
2.3	Light	12	1-2	0.42 ± 0.05	0.46 ± 0.04	0.40 ± 0.02	0.39 ± 0.03
4.4	Light-Moderate	3	2	0.45 ± 0.06	0.63 ± 0.10	0.32 ± 0.03	0.35 ± 0.02
18.1	Moderate-Heavy	12	0	0.53 ± 0.08	0.51 ± 0.06	0.40 ± 0.02	0.39 ± 0.04

For any parameter at a certain trial there were not significant differences before and after the presence of rain ($p>0.05$).
* A wastewater cycle applied to a bed provides approximately 0.006 mg N g$_{media}$$^{-1}$ and 0.0007 mg P g$_{media}$$^{-1}$.

8.3.5. Material quantities

Currently, after 2 years of operation, the sand is the material with the highest weight and volume in the CWR (73.5 and 45.2% of the total of a CWR 10 × 10 cm sample, respectively) (Column 4 and 5, Table 8.3). The roots and the organic soil, in the 10 × 10 cm CWR sample, had the lowest total volume percentage (11.6 and 12.5 %, respectively) among all CWR components, but they did not differ much from that of the LECA and PLA (Table 8.3).

In order to show the time changes of the material quantities, the volume of the original materials in the CWR was calculated using the design values (Column 3, Table 8.3). The volume of sand, LECA, PLA beads and organic soil in the original design was similar to that after 2 years of operation; while the original volume of roots was 4.6 times lower than the present volume (Column 3 vs. 5, Table 8.3).

8.4. DISCUSSION

8.4.1. Water movement and retention

The average moisture content along the bed length (424-388 mV) demonstrated that, in general, the whole bed received water. The moisture content variations encountered during a dry and wet period did not allow identifying the time taken by the water to flow along the length of the bed. Thus, the actual HRT was not well defined. See CWR pictures during dry and wet periods in Appendix E. Low effluent volumes for water reuse correlated to hot summer days where the wastewater was (almost) completely evapotranspirated close to the inlet area (Fig. 8.2A), due to the shallow depth and large area, turning the system often into a zero-discharge CW (Białowiec et al., 2014). This caused serious plant drought, deteriorating the CWR aesthetics in such period (Appendix E). Thus, a higher HLR that keeps the bed wet (and active) and provides effluent water for reuse is necessary. It should be noted that the CWR was designed to treat > 700 L d^{-1} (Zapater-Pereyra et al., 2013; Chapter 7) expecting a company growth, and currently only 499 L d^{-1} (Table 8.1) is being pumped. In addition, a lower roof inclination than that of the CWR (14.3°) would be beneficial as well, since the gravitational potential would have less influence on the water movement. During rainy days, the excess of water accelerated the wastewater movement along the length of the CWR, thus highly reducing the system's HRT (Fig. 8.2B-C). Water flows faster when soil pores are filled with water (saturated) than when they are filled with air (unsaturated), due to the reduction of the soil matric (suction) potential (Gallage et al., 2013). Nevertheless, the rain highly dilutes any pollutant and, in terms of concentration, the water quality was not negatively affected.

The CWR retained approximately 62% of the influent water (Table 8.1). This value is similar to that shown by Berndtsson (2010) for heavy rain (63%) in GRs. This is of particular importance in cities during rain events to attenuate the amount of water runoff reaching (and often collapsing) the sewer system and treatment plants.

8.4.2. Aerobic characteristics of the CWR

Oxygen provided via atmospheric diffusion is known to be an order of magnitude lower than that needed in a conventional CW treating domestic wastewater (Nivala et al., 2013). Tanner and Kadlec (2003) reported values of ~0.11 g O_2 m^{-2} d^{-1} that can be provided to the water surface of a conventional subsurface flow CW. This value can be slightly higher due to some space created by roots or the movement of macroinvertebrates. Nevertheless, it is unlikely to be higher than 1.0 g O_2 m^{-2} d^{-1} (Tanner and Kadlec, 2003). Using the equation of Kadlec and Wallace (2009) to calculate the oxygen consumption rate (OCR) (NH_4^+-N was replaced by TN), the maximum areal OCR in the CWR was ~2.3 g O_2 m^{-2} d^{-1} (Table 8.6). This value is much higher than that supposedly provided via atmospheric diffusion.

Plants provide 0-12 g O_2 m^{-2} d^{-1} to conventional CWs (Kadlec and Wallace, 2009). The roots in the CWR occupy ~12% of the whole media (Table 8.3), while in conventional CWs they occupy ~1-2% (unpublished data from a ~1.3 yr old vertical flow CW laboratory setup). Thus, the grass could provide a great part of the oxygen used by the CWR. Furthermore, a shallow bed favors the atmospheric diffusion of oxygen (Albuquerque et al., 2009) and results in a high root density that contributed to a higher water-root contact (García et al., 2005) than that occurring in deep beds, hence increasing the chances for oxygen release from the roots to the water. The high amount of (highly oxygenated) rain water (55% of total input water, Table 8.1) creates mixing in the bed that can as well facilitate oxygen transfer (Kadlec and Wallace, 2009).

8.4.3. CWR performance

The shallow characteristic of the CWR bed is the foremost cause of the highly aerobic system and the consequent improved performance (removal efficiency of > 79% for COD, > 95% for BOD_5 and > 99.7% for NH_4^+-N, Table 8.2). This is in agreement with García et al. (2005) who found a better performance of shallow CWs (27 cm water depth) than in conventional CWs (50 cm water depth). Despite the limitations of the PNR test to predict maximum the maximum potential capacity (see Section 5.4.4), the actual nitrification rate (0.04 mg N L^{-1}_{media} h^{-1}) was lower than the PNR (0.2 mg N L^{-1}_{media} h^{-1}), suggesting that the system nitrification efficiency can go further and probably other aerobic treatment processes as well (e.g. aerobic organic matter degradation). This is consistent with the fact that a low OLR was applied to the CWR (to avoid clogging) (Table 8.6) allowing the (almost) complete removal of NH_4^+-N and BOD_5, therefore the oxygen available did not fully deplete (effluent DO > 5 mg L^{-1}). Thus, considering oxygen availability, the CWR is able to cope with a higher OLR (or HLR).

Denitrification took also place in the oxic CWR (Table 8.2). NO_3^--N adsorption onto LECA is considered to be negligible (Albuquerque et al., 2009). A possible explanation might be that, despite the highly aerobic conditions, the deeper layers of the biofilm became anoxic (Stewart, 2003), where heterotrophic denitrifiers can develop. However, the possibility of TN removed by other removal pathways such as aerobic denitrification (Robertson and Kuenen, 1984) should not be discarded. Aerobic denitrification (direct transformation of NH_4^+-N to N_2

or NO_x species without the production of NO_2^--N or NO_3^--N) is a process receiving increasing attention (Kadlec and Wallace, 2009), also in CWs research (Shipin et al., 2005; Austin et al., 2006; Wang et al., 2009; Ji et al., 2012). Aerobic denitrifiers are present in a high number in natural soil samples (Kadlec and Wallace, 2009), thus their presence in CWs cannot be excluded as, for instance, *Paracoccus spp.* was found in a free water surface CW (Shipin et al., 2005) and in a tidal flow CW (Austin et al., 2006). Probably the micro-porosity of LECA enhanced a quick and stable biofilm development for both autotrophic and heterotrophic (Albuquerque et al., 2009). The PDR suggests that a much higher amount of nitrogen could be heterotrophically denitrified compared to the actual denitrification rates (Fig. 8.4) if ideal anoxic conditions would occur in the CWR. However, this value cannot be taken as a real PDR as the CWR is a highly aerobic system, thus such anoxic conditions will not occur.

Table 8.6 Hydraulic loading rates (HLR), organic loading rates (OLR) and oxygen consumption rates (OCR) in the constructed wetroof (CWR) and other studies.

	Depth (cm)	HLR (cm d^{-1})	OLR		OCR		Reference
			Areal (g BOD$_5$ m^{-2} d^{-1} / g COD m^{-2} d^{-1})	Volumetric (g BOD$_5$ m^{-3} d^{-1} / g COD m^{-3} d^{-1})	Areal (g O$_2$ m^{-2} d^{-1})	Volumetric (g O$_2$ m^{-3} d^{-1})	
CWR	9	0.2	0.4 / 1.2	3.9 / 12.2	2.3	22.3	This study
GROW	16	6.8	11.3 / 33.9	70.3 / 212.1	NA	NA	Frazer-Williams et al. 2006; Winward et al. 2008
HF CW	25	1.8	4.2 / NA	16.8 / NA	7.9	31.6	Nivala et al. 2013
HF CW	50	3.6	8.3 / NA	16.6 / NA	12.9	25.8	Nivala et al. 2013
VF CW	85	9.5	21.9 / NA	25.8 / NA	49.8-58.6	58.6-68.9	Nivala et al. 2013
HF CW + aeration	100	13.0	31.0 / NA	31.0 / NA	87.5	87.5	Nivala et al. 2013
VF CW + aeration	85	9.5	22.2 / NA	26.1 / NA	62.3	73.3	Nivala et al. 2013
Reciprocating	95	16.0	30.1 / NA	31.7 / NA	84.8	89.3	Nivala et al. 2013

GROW, Green roof water recycling system; HF CW, horizontal flow constructed wetland; VF CW, vertical flow constructed wetland
NA, not available.
Depth for Nivala et al. (2013) is indicated as effective depth. Values shown in this table refer to the planted systems of that manuscript.

For phosphorus, adsorption onto LECA seems to be the reason for the high removal efficiency (> 86%, Table 8.2). This material is known for its phosphorus removal properties (Vohla et al., 2011; Yaghi and Hartikainen, 2013), with a removal capacity ranging from 0 to 12 mg P g^{-1} (Vohla et al., 2011), although some phosphorus desorption from LECA has been reported in the literature (Yaghi and Hartikainen, 2013). For comparison, gravel and sand showed values of 3-3.6 and 0.27-3.9 mg P g^{-1}, respectively (Vohla et al., 2011). Assuming an adsorption capacity of 12 mg P g^{-1}, all the LECA in the CWR should be able to adsorb in total ~23 kg P and be exhausted after ~4.2 yr since construction. Plant uptake was neglected since the harvested grass was not removed from the CWR surface (see Section 8.4.4). Precipitation as hydroxyapatite might have been a phosphorus removal mechanism as well. The water pH and the high EC removal (probably due to a high loss of Ca^{2+} ions) supported that hypothesis (Table 8.2). This is based on the fact that the EC is a relatively stable parameter in CWs (Kadlec and Wallace, 2009) and the decrease in its concentration is only associated to rain dilution. But its final concentration was lower than what the rain alone could dilute.

It should be noted that rain can also dilute the pollutants load. However, from the water balance (Table 8.1), the dilution caused by the rain (and the sprinkler water) is only 1.4 times. If the achieved treatment was only the effect of dilution by rain, concentrations would have been much higher than those shown in Table 8.2. For instance, TN would have been 179 mg L^{-1} instead of the observed 19-32 mg L^{-1} (Table 8.2). Thus, dilution by the rain water cannot be taken as a major removal mechanism in the CWR. On the contrary, the effect of rain was counteracted by the losses of water via evapotranspiration (Table 8.1) that tends to concentrate pollutants in the water due to the water loss (Białowiec et al., 2014).

Comparison between the CWR and conventional CWs should be done in terms of volume and not area (see Section 8.4.6). Using the volumetric OCR for comparison among systems, the CWR maximum volumetric OCR was in the range of that in conventional HF CWs, while much lower (3-4 times) than those in vertical flow and intensified CWs (Table 8.6). Due to the oxic characteristic, a higher OCR was expected in the CWR. Probably this occurred due to the low OLR applied (1.2 g COD m^{-2} d^{-1}) intended to avoid clogging and thus to extend the life span of the system. Some studies recommend < 25 g COD m^{-2} d^{-1} (Kadlec and Wallace, 2009) to avoid clogging in conventional CWs, thus the OLR in the CWR could increase.

The most similar system (on a roof) reported in the literature, called GROW (green roof water recycling system; Avery et al., 2006; Frazer-Williams et al., 2006; Winward et al., 2008), treated a continuous flow of greywater (7 cm d^{-1}, Frazer-Williams et al., 2006) at a higher areal and volumetric OLR than the CWR investigated (Table 8.6), but with some particular advantages: 16 cm depth, not exposed to the environment (membrane covered) and includes artificial aeration. The CWR achieved, nevertheless, a better BOD_5 and similar COD and TSS effluent concentrations to that in the GROW system.

Furthermore, the CWR effluent was capable to meet up to the strictest level (Class IIIb, in mg L^{-1}: 60 for TSS, 40 for BOD$_5$, 200 for COD, 4 for NH$_4^+$-N, 60 for TN and 6 for TP) of a local guideline for discharge into water bodies (issued by the Commissie Integraal Waterbeheer; CIW, 1999), since the overflow water from the CWR was sent to an infiltration pond. Guidelines that include water reuse for toilet flushing in other parts of the world (e.g. USEPA, 2012; Health Canada, 2010) are stricter, for organic matter and solids (in mg L^{-1}: 20 for TSS and 10-20 for BOD$_5$), than the local discharging guideline and in addition, they include microbial parameters. The CWR meets their organic matter and solids requirements as well. Measurement of fecal indicator organisms was not part of this study and should be further investigated.

8.4.4. Media nutrient retention capacity

The nutrient balance showed clearly that 85% and 46% of nitrogen and phosphorus, respectively, were found neither in the effluent water nor in the media (Table 8.4). It is worth noting that grass uptake was not included in this balance as the harvested grass was not actively removed from the surface, thus a majority of the nutrients was expected to return to the CWR. A little part of the grass might have been removed due to the wind.

Nitrogen losses were due to the high denitrification encountered. The phosphorus losses might have been the result of precipitation and settling at the bottom of the CWR, where it was difficult to sample and therefore not considered in the balance calculations. Also, it was not possible to measure the original turf mat nutrient content, thus it was not subtracted in the calculation and as a consequence, the influent nutrient content is higher than what it could have been. Probably that difference could explain the phosphorus losses as well.

The CWR showed to be a stable system during a rain event. The nutrients retained by the media (3.8 kg N yr^{-1} and 2.8 kg P yr^{-1}, Table 8.4) were not released again in the water after a rain event of up to moderate-heavy category (Table 8.5). Thus, rain did not impact in the effluent quality by washing of nutrients immobilized on the media. It should be noted that these experiments were conducted after a short period of rain. Longer periods of rain with the same intensity would probably have resulted in a different outcome. However, the larger the amount of rain, the stronger the dilution effect.

8.4.5. Media contribution to the treatment

The media were analyzed by the FDA assay, before construction ("original") and after 2 years of operation ("present") to investigate in which material the major activity for the treatment took place (Table 8.3). The original media confirmed that no activity occurred previous to wastewater applications (Column 1, Table 8.3), thus none of the activity encountered in the present material could be related to an existing microbial activity. The difference in the microbial activity between the original and present materials (Columns 1 and 2, respectively, Table 8.3) showed a high development of bacterial biomass after 2 years of operation, as FDA hydrolysis correlates with microbial biomass (Sánchez-Monedero et al., 2008).

When comparing the turf mat vs. the substratum, the microbial activity in the former is much higher than that in the latter (Fig. 8.3). The turf mat is composed of organic soil and roots, both support an active biofilm. The turf mat is comparable to soils amended with compost as similar activities (100-300 $\mu g_{fluorescein}$ g^{-1} h^{-1}) were found in that material (Sánchez-Monedero et al., 2008) as in the turf mat (230-330 $\mu g_{fluorescein}$ $g^{-1}_{material}$ h^{-1}). Point 1 showed the higher microbial activity as compared to the rest of the bed (Fig. 8.3). This can be explained by its proximity to the inlet area where the highest organic loading occurs (Kadlec and Wallace, 2009) and the fact that, during some sunny days, it was the only point receiving domestic wastewater as it would completely evaporate before reaching Point 2. This higher activity in Point 1 explains the high PNR in Point 1 as well (Fig. 8.4).

The LECA, and especially the PLA, did not seem to contribute significantly to the microbial treatment as compared to the other CWR components (Column 2, Table 8.3). However, both materials were used for their light weight properties as the LBC of the building was only 100 kg m^{-2} and at least 9 cm bed depth had to be achieved (Zapater-Pereyra et al., 2013 - Chapter 7). Furthermore, LECA showed its benefits for phosphorus removal. The microbial activity was by far higher in the turf mat and roots than in the sand (Column 2, Table 8.3). The oxic micro zones in the rhizomes, the exudates excreted by the roots and the high amount of carbon and nutrients, created an optimal location for bacteria in the turf mat. Nonetheless, the amount of sand in the CWR is more than 3.6 times the amount of the other materials (Column 4 and 5, Table 8.3). So, in absolute terms for the whole CWR (in $g_{fluorescein}$ h^{-1}), the organic soil provided the majority of the activity, followed by the sand, roots, LECA and PLA (Column 6, Table 8.3). The absolute contribution of the roots to the treatment is considered high as compared to other CWs, since the CWR roots take a larger part of the total volume (~12%) compared to conventional CWs (~1-2%).

The microbial activity results highlight the importance of the turf mat in the CWR design. It provided an immediate ideal condition for biofilm development until the system has adapted, apart from the initial purpose of providing an immediate green aspect and coverage to protect the substratum from external factors (e.g. rain, snow, wind, maintenance people and animals) (Zapater-Pereyra et al., 2013 - Chapter 7), until the few roots (2.5% of the original CWR volume) have grown and established to ~12% after 2 years of operation (Column 3 and 4, Table 8.3).

8.4.6. **Area and volume requirements by the constructed wetroof**

Using the data obtained in this study (217 mg BOD_5 L^{-1} and 499 L d^{-1} of wastewater, Table 8.2 and 8.1, respectively) and considering that 1 PE (population equivalent) = 60 g BOD_5 d^{-1} (Eq. 2.1), the area used by the CWR was calculated to be ~170 m^2 PE^{-1}. This value seems to be very high when compared to the space taken by conventional CWs (~1-5 m^2 PE^{-1} for the removal of organics and TSS, Vymazal 2011). However, due to the CWR location, a commonly unexploited space (roof), the (land) area taken can be considered as 0 m^2 PE^{-1}.

Therefore, comparing the CWR in areal terms with conventional CWs should be avoided and volumetric comparison is more suitable (Table 8.6), since the CWR was designed based on a conventional CW volume divided by the roof area (306 m^2).

A higher LBC of a building could definitely hold a deeper CWR (higher volume), thus it could receive more wastewater and reduce the area demand. An example is the Water Tower project in 2010 (Bussum, The Netherlands), where a conventional CW of 1.15 m depth was built on the roof of a technical room with an LBC of ~3 ton m^{-2} (www.ecofyt.nl). This shows that the CWR is a case dependant system with a flexible design depending on the LBC of the building.

8.5. CONCLUSIONS

▪ The CWR is a highly aerobic system that treats domestic wastewater further beyond the levels required by local water quality guidelines for organic matter, solids, NH_4^+-N, total nitrogen and total phosphorus.

▪ The roots, organic soil and the sand provided the majority of the treatment. LECA and PLA contributed the most for the CWR volume (and weight) requirements. LECA also played a significant role in phosphorus removal.

▪ Rain has a minor effect in the treatment and no effect on washing away the nutrients retained in the CWR media. It, however, influenced the water movement along the bed length.

▪ During sunny days the wastewater is completely evaporated close to the inlet, thus the water never reached the outlet. The majority of the pollutants concentrate close to the inlet area providing the conditions for a high microbial activity.

▪ This CWR had an area of ~170 m^2 PE^{-1}, however the CWR design is flexible and can take less space depending on the LBC of the building. As the CWR is placed on a roof it can be considered as 0 m^2 PE^{-1}.

CHAPTER 9.
GENERAL DISCUSSION AND CONCLUSIONS

9.1. INTRODUCTION

Records of the raising population in the world bring dire consequences to the water sector such as the increment of fresh water demand and the subsequent decrease of fresh water sources, the massive wastewater production that demands more sanitation technologies and the dwindling of (green) areas due to extensive construction. Natural wastewater treatment systems such as constructed wetlands (CWs) are able to counteract such consequences since they are efficient and low-cost wastewater treatment technologies with a high potential for integration into the urban surroundings following the concept of water sensitive urban design (Section 1.1, Wong, 2006; Eisenberg et al., 2014).

CWs occupy large areas to guarantee an efficient treatment, usually much larger than other compact energy-intensive wastewater treatment systems (Table 2.1). This characteristic can become a major concern when land availability is scarce, e.g. in big cities and megacities or in mountain regions where large flat land surfaces are rarely found (Kadlec and Wallace, 2009; Foladori et al., 2012) and can be the reason for not being fully promoted (Wu et al., 2014). Compact wastewater treatment systems were and are necessary. However, the majority of compact systems namely biofilter, activated sludge, biological aerated filter and biological fluidized beds are energy intensive and none are able to integrate well in the surroundings (Eisenberg et al., 2014).

All treatment technologies have advantages and drawbacks; thus, the selection of any technology relies on two factors: their services and costs. Some services can be, for example, simplicity, environmental friendly, guarantying a certain effluent quality, providing an aesthetic value, able to be integrated within the urban surroundings (e.g. park) and/or being compact. The second factor includes the capital cost (land, design and construction) and the operation and maintenance costs. Usually those costs (excluding land cost) are larger in compact systems (e.g. activated sludge) than in non-energy intensive systems (Arceivala and Asolekar, 2007). Achieving many services at a low cost is difficult using conventional wastewater treatment systems. For example, using an activated sludge guarantees a compact, but not an aesthetic system that can be integrated within the urban surroundings if necessary. A conventional CW guarantees a green and aesthetic value but demands larger areas. Therefore, this thesis designed and studied two natural systems, i.e. the Duplex-CW and the constructed wetroof (CWR), in order to develop simple wastewater systems with a low space requirement, little energy consumption and a great capability of integration into the surroundings.

9.2. DESIGN AREAS OF CONSTRUCTED WETLANDS

It is known that conventional CWs require large areas for their construction. CWs are classified as natural wastewater treatment systems and as such, they only employ mechanisms found in nature to improve the quality of the applied wastewater, without the use of energy. Quoting Arceivala and Asolekar (2007): "...nature is slower, requiring longer detention periods which in turn imply larger land elements...". Many studies have mentioned that CWs require large areas, but not all have presented that as a disadvantage since not always there is a lack of area. Nevertheless, with the rapid population growth, areas are getting scarcer and in many places demand is already higher than the supply (e.g. megacities).

Basically the use of natural wastewater treatment systems is a trade-off: either more (money for) land or energy-consuming equipment (Mara, 2006). CWs have not been the exception for this trade-off. An overview of the CW literature since their spread (in the 80's) is presented in Chapter 2. The immediate intention of the majority of the investigations has been, throughout the years, to improve the CWs performance. In other words, increasing the treatment efficiency of each portion of CW and indirectly, achieving a lower CW footprint. Different strategies have been adopted to boost the CW performance, e.g. use of energy for recirculation, aeration or tidal operation, use of larger resting periods and use of reactive media for phosphorus removal (Section 2.1). Other newer strategies employed, to a lesser extent, to decrease the area are to stack extra treatment stages (Section 2.2) and to place the systems at unused spaces (Section 2.3).

For example, among the 20 VF CW studies (alone or in combination with HF CWs, excluding the green walls) summarized in Table 2.2 (including tidal flow CWs), the design land areas were in the range of 0.01-2.3 m^2 PE^{-1}. The lowest area corresponds to a tidal flow CW (Zhao et al., 2004b,c) and the largest to a combination of a HF and VF CW in a towery design (stacked arrangement) that included cascades for passive aeration (Ye and Li, 2009). This range clearly shows that currently only energy-intensive CWs (e.g. tidal CW) can reach very little area, just as or below any other energy-intensive technology (Table 2.1), while areas of about 2 m^2 PE^{-1} are still possible to reach without energy requirements such as the towery design of Ye and Li (2009) or the "French systems" (two stages of VF CWs treating raw wastewater developed in France) (Troesch et al., 2014).

9.3. APPROACHES IN THIS THESIS TO SELECT THE REQUIRED CONSTRUCTED WETLAND AREA

Along the years, different methodologies to design CWs have been developed evolving from rule of thumb to regression equations, plug-flow first order models, Monod-type models and compartmental models (Rousseau et al., 2004; Mburu et al., 2013). In the majority of the cases, they result in a different CW design, which complicates the comparison of the area demand of the systems. To avoid this, all comparisons among technologies in this thesis were done in terms of m^2 per population equivalent (PE). Literature usually reports this value, however, when it was not reported, the PE value was calculated using Equation 2.1. Note that such values should be interpreted as the "design area" demand and do not guarantee a successful performance of the CW. The area taken by the system that guarantees the treatment success, from now on referred to as "required area", should involve effluent concentrations. For that, two approaches were used in this study: the *guidelines* and the *first order model* approach.

The first approach, applied to an individual system (Chapter 4-6 and 8), verified if the effluent concentration met certain quality *guidelines* and only if it did, the "design area" of a specific system could be taken as the "required area". The second approach used a *first order model* (assuming no background concentration) to obtain the space that a system could save as compared to another reference system. This second approach was used for the Duplex-CWs (Chapter 4-6) despite accuracy might be low since first order models assume constant conditions (e.g. influent, flow and concentrations) and ideal plug flow behavior (Rousseau et al., 2004). Many researchers and practitioners have, however, used them for the design of other types of CWs, e.g. vertical flow (VF) with intermittent feeding (Kantawanichkul et al., 2003; Kadlec and Wallace, 2009), therefore it was used in this thesis as well (but only for comparison reasons).

9.4. OPTIMIZATION OF THE PERFORMANCE AND AREA REQUIREMENT OF THE DUPLEX-CW

HF CWs commonly require areas of 5 m^2 PE^{-1} (Vymazal, 2011) to remove organics and suspended solids. To remove nitrogen (to < 8 mg L^{-1}) and phosphorus (to < 1.5 mg L^{-1}), larger areas of approximately 15-30 m^2 PE^{-1} and 40-70 m^2 PE^{-1}, respectively, are suggested (Schierup et al. 1990, cited in Babatunde et al. 2008). In an attempt to develop a compact CW able to remove organic matter, solids and nutrients in a single unit, different bench scale CWs were studied (Chapter 4-6). Various aspects were investigated (Section 9.4.1-9.4.5) with the intention of developing a compact CW.

9.4.1. Type of constructed wetland and arrangement

Two types of CWs (i.e. HF and VF) and arrangements (i.e. conventional and stacked) were used in this study. The (continuous flow) HF CWs in a conventional (Control and Aerated) and stacked arrangement (Hybrid) (Chapter 4) did not show satisfactory nitrogen removal (0-40%, Table 4.1). In the Control, the low oxygen concentration hampered nitrification and in the Aerated and Hybrid systems, the excess of oxygen and the excess of bypass water (see Section 9.4.4) limited denitrification, respectively. Thus, a stack of a VF CW over a HF filter (HFF) was investigated (the Duplex-CW, Chapter 5-6). VF CWs usually require less area (1-3 m^2 PE^{-1}) than HF CWs (Vymazal, 2011) due to the intermittent loadings that allow resting periods, essential to enhance aerobic degradation and nitrification (Kadlec and Wallace, 2009). The added value of a CW with a stacked arrangement (like the Duplex-CW) is the combination of different technologies in a single unit, enhancing removal of more pollutants without compromising the area.

9.4.2. Water flow operation and configurations

Continuous flow in HF CWs (Chapter 4) and batch flow in the Duplex-CWs (Chapter 5 and 6) were studied and three configurations (modes of operation) were tested for the Duplex-CWs (Chapter 6): fill and drain system (Fill&D), stagnant batch (StagB) and free drain (FreeD) (Section 6.2.1). The Fill&D Duplex-CW outperformed all the other tested systems

(Table 4.1, Figs. 5.3 and 6.3-6.4). The batch flow coupled with the hydraulic retention time (HRT, 1 d) of the VF CW compartment enhanced the bed oxygenation and provided more time to achieve better COD and NH_4^+-N removal efficiencies than in the other tested systems (Green et al., 1998; Ghosh and Gopal, 2010; Weerakoon et al., 2013). The saturated HFF (anoxic) removed the majority of the total nitrogen (TN) (Table 6.3).

9.4.3. Intensification

Artificial aeration and effluent recirculation were evaluated in this thesis to boost the Duplex-CWs performance (Chapters 4-6). Artificial aeration in the HF CWs (organic loading rates (OLR) of up to 20 g COD m^{-2} d^{-1}) improved the systems performance (Chapter 4), while in the Duplex-CW it did not play a major role up to 37 g COD m^{-2} d^{-1} (Chapter 5 and 6). The VF CWs of the Duplex-CWs operated with a batch flow that allowed for oxygenation of the bed, contrary to the HF CWs operation (Fig. 6.3 vs. Table 4.1). The intrinsic oxic condition of the VF CWs was enough to treat the applied OLRs, since the oxygen demand to treat the highest wastewater strength (30 g O_2 m^{-2} d^{-1}) was close to the oxygen transfer rate usually provided by conventional VF CWs (3.5-24.7 g O_2 m^{-2} d^{-1}, Kadlec and Wallace, 2009). Furthermore, low air bubble dispersion throughout the bed was visualized probably due to the low voids within the sand (Section 5.4.1, Appendix E). Similarly, the use of recirculation (Chapter 5) did not enhance the treatment performance for any of the tested parameters, e.g. organic matter, solids, pathogens and nitrogen (Fig. 5.3). In fact, it showed lower TN removal efficiencies than the Aerated and Control systems due to the slightly higher dissolved oxygen (DO) created by the movement of the recirculating water (Fig. 5.3) (Sun et al., 2003; Foladori et al., 2014). Therefore, intensification is not recommended in the Duplex-CW design for the applied hydraulic and organic loads.

9.4.4. Carbon as electron donor for denitrification

The hybrid stacked system (Chapter 4) and the Duplex-CWs (Chapter 5 and 6) were designed to enhance TN removal based on the combination of an aerobic upper compartment and an anoxic bottom compartment. Since the majority of the organic matter was degraded in the upper compartment, there was probably a lack of carbon (electron donor) for denitrification in the HFF. Therefore, 40% of the raw wastewater (influent) was bypassed to the HFF of the Hybrid CW (Chapter 4) and 3.5 L of filtered influent was applied to the HFF of the Duplex CW (Chapter 5). In none of the cases, the addition of a carbon source contributed to the TN removal. For the Hybrid CW, the excess of raw water bypassed carrying NH_4^+-N restricted nitrification, due to the reduced conditions (0.9 mg O_2 L^{-1}, Table 4.1) and the short HRT (1 d). Nitrification is significantly limited below 2 mg O_2 L^{-1} (Hammer and Knight, 1994) and it is usually slower than denitrification (Verhoeven and Meuleman, 1999). In the Duplex-CW, the manipulation and application of the filtered influent oxygenated the water increasing the DO concentrations in the HFF to 1.7-5.0 mg L^{-1} (Table 5.3), hampering denitrification since it cannot occur above 0.3-1.5 mg O_2 L^{-1} (Kadlec and Wallace, 2009). It was concluded that the use of influent as electron donors for denitrification is not necessary in the Duplex-CW design. Instead, enhancing anoxic conditions in the HFF becomes critical.

9.4.5. **Post-treatment for phosphorus removal**

CWs are usually not effective in removing phosphorus (Prochaska et al., 2007) and the tested CWs were not an exception (Table 9.1). Therefore, the phosphorus removal potential of oyster-shells, mussel-shells, crushed corals (raw and pyrolyzed) and nanoparticle-beads were tested in Chapter 3. Pyrolyzed oyster-shell beads and nanoparticle-beads were the most efficient materials, via precipitation and adsorption, respectively, removing >99% of the phosphorus. The high effluent pH (~12) after treatment with pyrolyzed material has a main limitation for discharge purposes, but is an advantage for disinfection (Parmar et al., 2001). The nanoparticle-beads maintain the water pH around neutral values. The nanoparticle-beads might have a large initial investment cost but the operational costs are highly reduced since the beads can be regenerated. The pyrolyzed oyster-shells need a low initial cost but a permanent production cost for pyrolysis and beads production. In addition, their use promotes the recycling of shells that are considered a waste in coastal areas and aquaculture (Lee et al., 2009). Both materials are recommended as a post-CW treatment.

9.4.6. **Area requirement**

The improvements made to the tested CWs (Sections 9.4.1-9.4.5) were done with the intention of reducing their area requirements. In Chapter 4, the two OLRs tested resulted in HF CW design areas of 6 and 11 m^2 PE^{-1}. None of the effluent concentrations, for both OLRs investigated, met the discharge guidelines for nutrients (Table 9.1). However, when the *first order model approach* was used, the results indicated that the use of the Aerated and the Hybrid systems instead of the Control system can save 1.5-50 times (depending on the parameter tested) the required area (Table 9.1). Nevertheless, the tested HF systems were not yet considered to be compact CWs since conventional CWs are designed with less than 6 m^2 PE^{-1}. Therefore, the CW design was modified from a HF CW-HFF (Chapter 4) to a VF CW-HFF (Chapter 5 and 6) and three higher wastewater loading rates (15, 27 and 37 g COD m^{-2} d^{-1}) were tested, resulting in design areas of 7.9, 3.4 and 2.6 m^2 PE^{-1} (Table 9.1, Fig. 9.1).

Following the *guidelines approach*, only the Fill&D Duplex-CW achieved 3.4 m^2 PE^{-1} (Table 9.1). According to those guidelines (Table 9.1), the same system reached an area requirement of 2.6 m^2 PE^{-1} for COD and TSS, but not for TN. However, the Fill&D met the TN removal efficiency of > 70% (in relation to the load of the influent) required by the same guidelines (EEC, 1991) and by the specifications used in Foladori et al. (2013). Thus, the Fill&D reached an area requirement of 2.6 m^2 PE^{-1} and saved 1.5-5 times the area that a single VF CW would have needed to remove TN (Table 9.1), according to the *first model approach*. For TP (Table 9.1), as mentioned in Section 9.4.5, a phosphorus removal post treatment is required. Although pathogen removal is not considered in the discharge guidelines (EEC, 1991), it should be noted that the Fill&D Duplex-CW (tested in Chapter 5) removed up to 5 logs of *E. coli* and fecal coliforms (Fig. 5.2), similar to removal values found in other CWs (Kadlec and Wallace, 2009; García et al., 2013). However, if the effluent is needed for agriculture (≤ 200 fecal coliforms per 100 mL, USEPA, 2012), the effluent concentrations would require further disinfection (e.g. ultraviolet radiation, Gross et al., 2007, 2008).

Table 9.1 Evaluation of the required area (*guidelines approach*) and area saved (*first order model approach*) of all tested constructed wetlands from Chapter 4, 5 and 6.

Characteristics	Parameter	Does it meet discharge guidelines?*			Can the area be reduced? How much?**		
CHAPTER 4		Control	Aerated	Hybrid	Control	Aerated	Hybrid
16.3 L d⁻¹	COD	✓	✓	✓	R.	✗	✗
11 g COD m⁻² d⁻¹	TSS	✓	✓	✓	R.	≥ 1.5	≥ 1.5
11 m² PE⁻¹	NH₄⁺-N	NA.	NA.	NA.	R.	≥ 10	≥ 5.0
	TN	✗	✗	✗	R.	✗	✗
	PO₄³⁻-P	✗	✗	✗	R.	✗	✗
16.3 L d⁻¹	COD	✓	✓	✓	R.	≥ 1.5	≥ 1.5
20 g COD m⁻² d⁻¹	TSS	✓	✓	✓	R.	✗	✗
6 m² PE⁻¹	NH₄⁺-N	NA.	NA.	NA.	R.	≥ 45	≥ 5.0
	TN	✗	✗	✗	R.	≥ 1.5	≥ 1.5
	PO₄³⁻-P	✗	✗	✗	R.	✗	✗
CHAPTER 5		Control	Aerated	Recirc.	Control	Aerated	Recirc.
11 L d⁻¹	COD	✓	✓	✓	R.	✗	✗
13 g COD m⁻² d⁻¹	BOD₅	✗	✗	✗	R.	✗	✗
8 m² PE⁻¹	TSS	✗	✓	✓	R.	✗	✗
	NH₄⁺-N	-	-	-	R.	✗	✗
	TN	✓	✓	✗	R.	✗	✗
CHAPTER 6		Fill&D	StagB	FreeD	Fill&D	StagB	FreeD
11 L d⁻¹	COD	✓	✓	✓	✗	✗	≥ 1.5
15 g COD m⁻² d⁻¹	TSS	✓	✓	✓	✗	≥ 1.5	≥ 1.5
7.9 m² PE⁻¹	NH₄⁺-N	NA.	NA.	NA.	✗	✗	✗
	TN	✓	✓	✗	≥ 1.5	✗	≥ 25
	TP	✓	✓	✓	≥ 1.5	≥ 1.5	≥ 1.5
11 L d⁻¹	COD	✓	✓	✓	✗	≥ 1.5	≥ 1.5
27 g COD m⁻² d⁻¹	TSS	✓	✓	✓	≥ 1.5	≥ 1.5	≥ 1.5
3.4 m² PE⁻¹	NH₄⁺-N	NA.	NA.	NA.	✗	✗	✗
	TN	✓	✗	✗	≥ 1.5	✗	≥ 5.0
	TP	✗	✗	✗	✗	≥ 5.0	✗
11 L d⁻¹	COD	✓	✓	✗	✗	✗	≥ 1.5
37 g COD m⁻² d⁻¹	TSS	✓	✓	✗	✗	✗	✗
2.6 m² PE⁻¹	NH₄⁺-N	NA.	NA.	NA.	✗	✗	✗
	TN	✗	✗	✗	≥ 1.5	✗	≥ 1.5
	TP	✗	✗	✗	✗	≥ 1.5	✗
11 L d⁻¹	COD	✓	✓	✓	✗	✗	≥ 1.5
37 g COD m⁻² d⁻¹	TSS	✓	✓	✓	✗	✗	≥ 1.5
2.6 m² PE⁻¹	NH₄⁺-N	NA.	NA.	NA.	✗	≥ 1.5	≥ 1.5
Aeration	TN	✗	✗	✗	≥ 1.5	✗	≥ 10
	TP	✗	✓	✗	✗	≥ 1.5	≥ 1.5

Recirc., Recirculating; Fill&D, fill and drain; StagB, stagnant batch; FreeD, free drain; NA., not available guidelines; R., reference point.

Guidelines approach (EEC, 1991) (in mg L⁻¹): 125 for chemical oxygen demand (COD), 25 for 5-days biochemical oxygen demand (BOD₅), 35 for total suspended solids (TSS), 15 for total nitrogen (TN) and 2 for total phosphorus (TP).

**First order model approach*. The area of each system was reduced a certain amount of times as compared to a reference point (R.): ≥ 1.5, ≥ 5, ≥ 10, ≥ 25 and ≥ 45 (up to 50) times reduction. Reference points: Chapter 4, the Control HF CW; Chapter 5, the Control Duplex-CW and Chapter 6, their own VF CW compartment alone.

BOD₅ was not measured in Chapter 4 and 6.

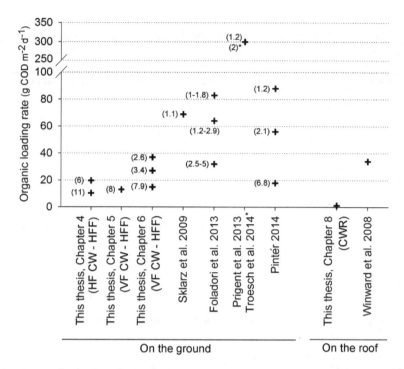

Figure 9.1 Comparison of applied organic loading rates used in this study to those from the literature (only vertical flow constructed wetlands, VF CWs). Values in brackets are the design areas, in m² PE⁻¹.

Systems description: Sklarz et al. (2009), recirculation VF CW; Foladori et al. (2013), recirculation VF CW; Prigent et al. (2013), VF CW filled with Mayennite, aeration pipes and fed raw wastewater; Troesch et al. (2014), classical French system with 2 stages and fed with raw wastewater; Pintér (2014), Duplex-CW with recirculation and Winward et al. (2008), CW on a roof covered with membrane to avoid rain and artificial aeration.

If less area than 2.6 m² PE⁻¹ is targeted, a higher HLR and/or OLR should be applied. Frequent intermittent feeding (15 min of pumping influent per hour, 5 days a week, 0.05-0.16 m³ m⁻² d⁻¹ and 13-88 g COD m⁻² d⁻¹) showed that the Duplex-CW is capable of reaching 1.2 m² PE⁻¹ for COD and TSS (based on the *guidelines approach*) (Pintér, 2014). The intermittent feeding (in the VF CW) and free drain accounted for an oxic environment that guaranteed a high COD (81-91%) and NH_4^+-N (61-99%) removal, but limited the TN removal (12-56%). Thus, TN guidelines were not met along that study and the design area of 1.2 m² PE⁻¹ is only valid if TN limits are not a requirement (Pintér, 2014). In this thesis, the TN removal achieved by the tested Duplex-CW was 39-87% and some of them were able to meet TN guidelines depending on the configuration and applied loads (Table 9.1).

Nevertheless, increasing the HLR (from 0.05 to 0.16 m³ m⁻² d⁻¹) and OLR (from 13 to 88 g COD m⁻² d⁻¹) triggered the benefits of recirculation (Pintér, 2014). The same effect was hypothesized in this study for artificial aeration, since in this study up to 37 g COD m^{-2} d^{-1} was applied and usually aerated systems are designed to treat higher loads (Table 5.4, Kadlec and Wallace, 2009). Due to technical limitations, testing higher OLRs was not possible in this thesis and further research is proposed (with and without the use of intensification) to identify the maximum capacity of the Duplex-CW. The research study of Kantawanichkul et al. (2001, 2003) and Kantawanichkul and Somprasert (2005) should be taken as a reference since they studied a similar system with high loads (105 g COD m^{-2} d^{-1}, 53-81 L d^{-1} for Kantawanichkul et al. 2001). Studies in Tables 2.2 and 5.4 and in Figure 9.1 should also be considered. The application of higher loads must be in any case done with caution, since bed clogging can occur rapidly (Knowles et al., 2011). Thus, in that case, aeration or long resting periods are highly recommended. For instance, a two stages VF CWs treating raw wastewater developed in France (called "French systems", Section 2.1.4) have always resting periods of 7 d (Troesch et al., 2014) and a compact French VF CW (one stage) works with resting periods of 7-14 d (Prigent et al., 2013). Further investigations should concentrate on determining the direct effect of oxygen (via aeration and/or resting periods of different lengths) on soil mineralization and clogging.

9.4.7. Duplex-CW area design in the literature context

CWs commonly need 3-5 m^2 PE^{-1}, though occasionally they are designed differently, e.g. < 0.1 m^2 PE^{-1} in Zhao et al. (2004a,b,c) and Sun et al. (2005); < 1.5 m^2 PE^{-1} in Ye and Li (2009) and Zhao et al. (2011); 7 m^2 PE^{-1} in Ouellet-Plamondon et al. (2006); 11 m^2 PE^{-1} in Jamieson et al. (2003) (Table 2.2). In this thesis, the Duplex-CW reached up to 2.6 m^2 PE^{-1}, without the use of energy, but showed a potential to reach lower areas. As shown in Section 9.2, designs already exist of around 2 m^2 PE^{-1} without the need of intensification. Therefore, regarding area, the Duplex-CW did not show significant improvement. Nevertheless, the concept of a stacked arrangement, not widely used, does help with land savings since, to reach further TN and pathogen removal, a HFF is needed and in a conventional way, it would have been placed next to the VF CW, taking the double area amount.

9.4.8. Recommended Duplex-CW design

Based on this thesis, the recommended Duplex-CW design consists of a Fill&D operation with a VF CW on top of a HFF, as described in Chapter 5 and 6, with an area requirement of 2.6-3.4 m^2 PE^{-1}. The system should include a post-treatment filter for phosphorus removal using nanoparticle-beads or pyrolyzed oyster-shells.

The Duplex-CW design requires no intensification (i.e. aeration and recirculation) up to the current operation conditions (up to 37 g COD m^{-2} d^{-1}, Fig. 9.1 and Table 9.1). If loads are increased and artificial aeration is included in the VF CW, a filter medium coarser than 1-2 mm is recommended for the appropriate distribution of air bubbles (Section 5.4.1). Since the HRT would be compromised with this practice, a valve should be installed to control the HRT in the VF CW, as in this study (Figs. 5.1 and 6.1).

9.5. DEVELOPMENT OF THE CONSTRUCTED WETROOF

A further improvement, from the area achieved by the Duplex-CW (Section 9.4.8) was intended by developing a CW on a roof (Chapter 7 and 8). The idea of designing a CW on a roof immediately implies a design area of 0 m^2 PE^{-1}, however only the confirmation of a successful performance would transform the design area into a required area.

9.5.1. Filter material selection and arrangement

Several trials with different materials (Table 7.3) were conducted to identify the appropriate material that could be used for the CWR filter medium (Chapter 7). The CWR media need to meet the load bearing capacity of the building (100 kg m^{-2}) to achieve an appropriate HRT (expected > 2 d, Table 7.2) and to guarantee the stability and long term life of the system. The combination of sand, polylactic acid beads (PLA) and light expanded clay aggregates (LECA) with embedded stabilization plates and covered by a turf mat met all the conditions and was used to build the pilot CWR studied in Chapter 8.

9.5.2. Aerobic conditions and effluent quality

The CWR demonstrated to be a robust wastewater treatment system able to efficiently treat domestic wastewater due to its high aerobic environment facilitated by the shallow 9-cm bed. This characteristic explains the enhanced BOD$_5$ (> 95%), COD (> 79%) and NH$_4^+$-N (> 99%) removal (Table 8.2). The CWR was also capable of a high TP (> 86%) and TN (> 87%) removal. The roots, organic soil (of the turf mat) and the sand provided the majority of the microbial treatment (Table 8.3). The phosphorus adsorption capacity of LECA (Vohla et al., 2011) explained the high TP removal in the CWR. For TN it was postulated that aerobic denitrifiers could have be present in the CWR as they have been found in other CWs (Shipin et al., 2005; Austin et al., 2006).

The success of the CWR performance and the "required" area achieved (0 m^2 PE^{-1}) was validated by the *guidelines approach* (Table 9.2). The CWR met the local guidelines for discharge and the international reuse guidelines (Table 9.2). European discharge guidelines for COD and TN concentrations were not met (Table 9.2) although the CWR effluent concentrations were very close to these guidelines (CWR/guidelines: 132/125 mg L^{-1} for COD and 19/15 mg L^{-1} for TN). However, both parameters met the minimum percentage of load reduction demanded in the same European guidelines (for COD 75% and for TN 70-80%; EEC, 1991). Pathogens are a relevant parameter for toilet flushing, however due to time constraints; these were not measured in this thesis. Therefore, the pathogen removal capacity of the CWR it is proposed for further investigation.

Table 9.2 Evaluation of the required area (*guidelines approach*) of the constructed wetroof (Chapter 8).

Characteristics	Parameter	Does it meet guidelines?*		
		Guideline I	Guideline II	Guideline III
	COD	✗	✓	-
499 L d^{-1}	BOD$_5$	✓	✓	✓
1.2 g COD m^{-2} d^{-1}	TSS	✓	✓	✓
0 m^2 PE^{-1}	NH$_4^+$-N	-	✓	-
	TN	✗	✓	-
	PO$_4^{3-}$-P	✓	✓	-

Guidelines approach: Guideline I, II and III refer to European discharge guideline (EEC, 1991), local discharge guideline (CIW, 1999) and international water reuse guideline for toilet flushing (Health Canada, 2010; USEPA, 2012). Guideline I (in mg L^{-1}): 125 for chemical oxygen demand (COD), 35 for total suspended solids (TSS), 15 for total nitrogen (TN) and 2 for total phosphorus (TP). Guideline II (strictest level, Class IIIb) (in mg L^{-1}): 60 for TSS, 40 for BOD$_5$, 200 for COD, 4 for NH$_4^+$-N, 60 for TN and 6 for TP. Guideline III (in mg L^{-1}): ≤ 20 for TSS, ≤ 10-20 for BOD$_5$.

9.5.3. Operation under different climate conditions: effect of temperature and rain

The performance of the CWR highly depended on the weather conditions, mainly temperature and rainfall intensity (Section 8.4.1). Temperatures above 25°C are optimal mainly for nitrogen removal (for nitrification 30°C (Gerardi, 2006) and for denitrification, 20-40°C (Sirivedhin and Gray, 2006)), but in a shallow and large (~170 m^2 of roof PE^{-1}) system like the CWR, they cause serious evapotranspiration leading to a limited plant growth (i.e. required for treatment and aesthetics; see CWR pictures in Appendix E) and a zero discharge system (Białowiec et al., 2014). Increasing the HLR of the system (during summer) is considered a potential solution to avoid plant damage and zero discharge and studying its effect on the CWR performance is proposed as further investigation. In fact, the CWR was designed for a higher HLR (~ 700-800 L d^{-1}, Table 7.2) projecting a growth of the building office, however during the study time, it only received 499 L d^{-1} (Table 8.1).

The CWR did not operate during winter (bed temperatures below 2°C) since it was hypothesized that the shallow bed and the large area would not prevent the freezing of the wastewater inside the system. Water freezing within the substrate occurred in ~20 cm depth green roofs (Teemusk and Mander, 2007). Conventional CWs during winter benefit from the temperature barrier created by the macrophytes and the underneath soil as a heat source (Kadlec and Wallace, 2009). The latter is not possible to occur in the CWR, however during winter the heating system of the building might create a temperature gradient underneath the CWR probably creating a heat flux. The potential benefits of such a heat flux and the plant isolation during winter should be investigated to fine-tune the CWR design.

Currently, the wastewater is discharged to the sewer system during winter. The periods of time with ambient temperatures < 2°C, vary depending on the country. In The Netherlands, it takes approximately 3-4 months per year with fluctuations during day and night, therefore discharging wastewater to the sewer system would be only during short periods of time. If such connection is not an option, a storage system should be required. However, since that might be difficult depending on the amount of water produced, the CWR treatment option should be limited to temperate climates. Alternatively, a deeper CWR bed might be necessary if the load bearing capacity of the building allows it. This would also guarantee larger HLR to be treated in case is needed. However, the CWR should remain relatively shallow to benefit from the aerobic treatment processes. The maximum depth of the CWR should be further investigated.

Rainfall showed different effects on the performance of the CWR. On one side, the amount of rainfall (Dutch annual rainfall of 700-900 mm, www.knmi.nl, Table 8.1) combined with the operational conditions and mild temperatures guaranteed a wet bed, which is beneficial in maintaining a microbial community (Sklarz et al., 2009) and for plants growth. On the other side, rainfall affected the CWR hydraulics by accelerating the wastewater movement along the wet bed and in consequence, reducing the system's HRT (Fig. 8.2B-C). Water flows faster when soil pores are filled with water (saturated) than when they are filled with air (unsaturated), due to the reduction of the soil matric (suction) potential (Gallage et al., 2013). This effect is even more pronounced when gravitational potential governs due to an inclined surface (the CWR slope = 14.3°, Table 7.2). Nevertheless, the CWR effluent quality was not deteriorated by the rapid water path since rain highly diluted the water (Section 8.4.3) and the nutrients retained in the CWR bed were not washed out up to Moderate-Heavy rain (Table 8.5, Section 8.4.4).

It can be inferred that the same CWR design with the current operation might not work in other places. For example, in equatorial areas (annual precipitation in the tropics usually exceeds 2500 mm), the wastewater would rapidly flow through the system compromising the HRT and the treatment performance. The treatment might become only a function of rain dilution without any treatment role of the CWR and probably, with more continuous rain nutrients will be washed away from the bed. Also, in dry and hot areas (e.g. Lima, Peru, < 15 mm yr^{-1}) might also be problematic to use CWRs since evapotranspiration will be very high and with the current operational conditions, the system might turn into a permanent zero discharge system. Further investigations should be conducted simulating weather scenarios different than that in The Netherlands to optimize the CWR design depending on the location.

9.5.4. Constructed wetroof in the literature context

In general, only energy intensive systems are able to have areas of < 1 m^2 PE^{-1} (Section 9.2, Chapter 2). The CWR requires 0 m^2 PE^{-1} and is thus considered the leading option to save area while guaranteeing treatment efficiency, simplicity, energy requirements and environmental benefits. Furthermore, the CWR is unique and no similar systems have been found in the literature. The most similar system still present various differences (e.g. bench scale study, high OLR and HLR applied continuously (Table 8.6, Fig. 9.1), 16 cm depth, use of artificial aeration, membrane covered to avoid exposure to environment; Winward et al., 2008). The CWR is the first real scale CW placed on a roof, exposed to all environmental factors (occurring in The Netherlands), with a successful performance since its construction in 2012.

The CWR becomes an important tool to treat wastewater in crowded megacities. However, since the CWR application requires a strong building structure, the likelihood for broad application (e.g. slum areas) will be reduced. The current design (9 cm depth) requires approximately 170 m^2 of roof PE^{-1} (Section 8.4.6), a relatively high value for a single household application. Therefore, that design is ideal for large buildings with large roof areas and low PE (e.g. shopping malls, office buildings). For the case of single households, either a more stable structural design (prior construction) of the house to increase the load bearing capacity and therefore to allow a deeper CWR matrix than that described in this study or a new type of urban planning where single households nearby large buildings with CWRs deliver their wastewater for treatment and subsequent recycling of water in common green areas are required.

9.5.5. Recommended constructed wetroof design

Since all the materials had an important function to guarantee the successful operation of the CWR and the chosen plants (grass) had also a major role in the treatment (see Section 8.4.5), the tested design of the CWR (as described in Section 8.2.1) is the recommended design. As the size of the CWR depends on the load bearing capacity of the building, this design should be taken as guidance for other projects. If the building structure allows, a deeper system that assures a larger HLR and diminishes the high evapotranspiration rates is recommended. However, the CWR should remain relatively shallow to benefit from the aerobic treatment processes.

9.6. CONCLUSIONS

▪ The Fill&D configuration outperformed all other tested CWs due to the oxygen operational conditions and the HRT. Therefore, it is the recommended Duplex-CW design which an area requirement of 2.6-3.4 m^2 PE^{-1}, except for TP. For almost all the cases, it met the required guidelines for organic matter, solids and nitrogen removal. For the removal of TP, nanoparticle-beads or oyster-shell beads can be used as post-treatment.

▪ The proposed Duplex-CW design does not require artificial aeration, recirculation nor the addition of an external carbon source (as electron donor for denitrification) to enhance TN removal. Artificial aeration only played an important role when HF CWs were used but not for the tested VF CWs. Probably higher OLRs would have triggered the benefits of artificial aeration and/or recirculation, resulting in an area demand below 2.6 m^2 PE^{-1}.

▪ The CWR is able to treat domestic wastewater during both sunny and rainy seasons occurring in The Netherlands (except winter) with a land requirement of 0 m^2 PE^{-1}. The CWR is a robust system highly recommended for domestic wastewater treatment as it can highly remove organic matter (> 95% for BOD$_5$ and > 79% for COD), nitrogen (> 87% for TN) and phosphorus (> 86% for TP).

REFERENCES

Abeynaike A., Wang L., Jones M.I., Patterson D.A. (2011). Pyrolysed powdered mussel shells for eutrophication control: effect of particle size and powder concentration on the mechanism and extent of phosphate removal. Asia-Pacific Journal of Chemical Engineering, 6, 231-243.

Abou-Elela S.I., Hellal M.S. (2012). Municipal wastewater treatment using vertical flow constructed wetlands planted with Canna, Phragmites and Cyprus. Ecological Engineering, 47, 209-213.

Adam G., Duncan H. (2001). Development of a sensitive and rapid method for the measurement of total microbial activity using fluorescein diacetate (FDA) in a range of soils. Soil Biology and Biochemistry, 33, 943-951.

Ahmed N. (2012). Using live panel to treat grey water. UNESCO-IHE MSc. Thesis ES12.24.

Albuquerque A., Arendacz M., Obarska-Pempkwiak H., Borges M., Correia M. (2008). Simultaneous removal of organic and solid matter and nitrogen in a SSHF constructed wetland in temperate mediterranean climate. KKU Research Journal, 13, 1-12.

Albuquerque A., Oliveira J., Semitela S., Amaral L. (2009). Influence of bed media characteristics on ammonia and nitrate removal in shallow horizontal subsurface flow constructed wetlands. Bioresource Technology, 100, 6269-6277.

APHA (1998). Standard methods for the examination of water and wastewater, 20th Edition, American Public Health Association, Washington, D.C., USA.

APHA (2005). Standard methods for the examination of water and wastewater. 21st Edition, American Public Health Association, Washington, D.C., USA.

APHA (2012). Standard methods for the examination of water and wastewater. 22nd Edition, American Public Health Association, Washington, D.C., USA.

Arias C.A., Del Bubba M., Brix H. (2001). Phosphorus removal by sands for use as media in subsurface flow constructed reed beds. Water Research, 35, 1159-1168.

Arias C.A., Brix H. (2005). Phosphorus removal in constructed wetlands: can suitable alternative media be identified? Water Science and Technology, 51 (9), 267-273.

Arceivala S.J., Asolekar S.R. (2007). Wastewater Treatment for Pollution Control and Reuse, Third Edition, McGraw-Hill, New Delhi, India, pp. 518.

Austin D.C., Lohan E. (2005). Patent: Tidal vertical flow wastewater treatment system and method. United States US 6,863,816 B2.

Austin D., Wolf L., Strous M. (2006). Mass transport and microbiological mechanisms of nitrification and denitrification in tidal flow constructed wetland systems. In: Proceedings of 10th International Conference on Wetland Systems for Water Pollution Control, Lisbon, Portugal. 23 - 29th September 2006, Vol. I, pp. 209-217.

Avery L.M., Frazer-Williams R.A.D., Winward G., Pidou M., Memon F.A., Liu S., Shirley-Smith C., Jefferson B. (2006). The role of constructed wetlands in urban grey water recycling. In: Proceedings of 10th International Conference on Wetland Systems for Water Pollution Control, Lisbon, Portugal. 23 - 29th September 2006, Vol I, pp. 423-434.

Ayaz SÇ., Aktaş Ö., Findik N., Akça L., Kinaci C. (2012). Effect of recirculation on nitrogen removal in a hybrid constructed wetland system. Ecological Engineering, 40, 1-5.

Babatunde A.O., Zhao Y.Q., O'Neill M., O'Sullivan B. (2008). Constructed wetlands for environmental pollution control: A review of developments, research and practice in Ireland. Environmental International, 34, 116-126.

Bäumer R. (2013). Waterzuivering van grijs afvalwater met de living wall. Howest (De Hogeschool West-Vlaanderen) MSc. Thesis.

Baveye P., Vandevivere P., Hoyle B.L., DeLeo P.C., Sanchez de Lozada D. (1998). Environmental impact and mechanisms of the biological clogging of saturated soils and aquifer materials. Critical Reviews in Environmental Science and Technology, 28, 123-191.

Bazargan S. (2014). Bio-treatment of domestic wastewater; using a living wall system. UNESCO-IHE MSc. Thesis 14.11.

Behrends L.L. (1999). Patent: Reciprocating subsurface-flow constructed wetlands for improving wastewater treatment. United States US 5,863,433.

Behrends L., Houke L., Bailey E., Jansen P., Brown D. (2001). Reciprocating constructed wetlands for treating industrial, municipal and agricultural wastewater. Water Science and Technology, 44 (11-12), 399-405.

Behrends L.L., Bailey E., Jansen P., Houke L., Smith S. (2007). Integrated constructed wetland systems: design, operation, and performance of low-cost decentralized wastewater treatment systems. Water Science and Technology, 55 (7), 155-161.

Berardi U., GhaffarianHoseini A., GhaffarianHoseini A. (2014). State-of-the-art analysis of the environmental benefits of green roofs. Applied Energy, 115, 411-428.

Berndtsson J.C. (2010). Green roof performance towards management of runoff water quantity and quality: A review. Ecological Engineering, 36, 351-360.

Białowiec A., Albuquerque A., Randerson P.F. (2014). The influence of evapotranspiration on vertical flow subsurface constructed wetland performance. Ecological Engineering, 67, 89-94.

Bianchini F., Hewage K. (2012). How "green" are the green roofs? Lifecycle analysis of green roof materials. Building and Environment, 48, 57-65.

Bitton G., Mitchell R., De Latour C., Maxwell E. (1974). Phosphate removal by magnetic filtration. Water Research, 8, 107-109.

Blaney L.M., Cinar S., SenGupta A.K. (2007). Hybrid anion exchanger for trace phosphate removal from water and wastewater. Water Research, 41, 1603-1613.

Boisen S., Bech-Andersen S., Eggum B.O. (1987). A critical view on the conversion factor 6.25 from total nitrogen to protein. Acta Agriculturae Scandinavica, 37, 299-304.

Boog J., Nivala J., Aubron T., Wallace S., van Afferden M., Müller R.A. (2014). Hydraulic characterization and optimization of total nitrogen removal in an aerated vertical subsurface flow treatment wetland. Bioresource Technology, 162, 166-174.

Bomo A.M., Stevik T.K., Hovi I., Hanssen J.F. (2004). Bacterial removal and protozoan grazing in biological sand filters. Journal of Environmental Quality, 33, 1041-1047.

Brix H. (1987a). Treatment of wastewater in the rhizosphere of wetland plants - The root-zone method. Water Science and Technology, 19, 107-118.

Brix H. (1987b). The applicability of the wastewater treatment plant in Othfresen as scientific documentation of the root-zone method. Water Science and Technology, 19 (10), 19-24.

Brix H., Johansen N.H. (1999). Treatment of domestic sewage in a two-stage constructed wetland—design principles. In: Nutrient Cycling and Retention in Natural and Constructed Wetlands, ed. J. Vymazal, Backhuys Publishers, Leiden, The Netherlands, pp. 155-163.

Brix H., Arias C.A. (2005). The use of vertical flow constructed wetlands for on-site treatment of domestic wastewater: New Danish guidelines. Ecological Engineering, 25, 491-500.

Butterworth E., Dotro G., Jones M., Richards A., Onunkwo P., Narroway Y., Jefferson B. (2013). Effect of artificial aeration on tertiary nitrification in a full-scale subsurface horizontal flow constructed wetland. Ecological Engineering, 54, 236-244.

Bykowski M.J., Ewing L. (1977). Patent: Phosphorus removal from wastewater. United States US 4,029,575.

Caselles-Osorio A., Puigagut J., Segú E., Vaello N., Granés F., García D., García J. (2007). Solids accumulation in six full-scale subsurface flow constructed wetlands. Water Research, 41, 1388-1398.

Chabaud S., Andres Y., Lakel A., Le Cloirec P. (2006). Bacteria removal in septic effluent: Influence of biofilm and protozoa. Water research, 40, 3109-3114.

Characklis W.G., Wilderer P.A. (1989). Structure and function of biofilms. John Wiley & Sons, Chichester, UK, 387 pp.

Chazarenc F., Gagnon V., Comeau Y., Brisson J. (2009). Effect of plant and artificial aeration on solids accumulation and biological activities in constructed wetlands. Ecological Engineering, 35, 1005-1010.

Choung Y.K., Jeon S.J. (2000). Phosphorus removal in domestic wastewater using anaerobic fixed beds packed with iron contactors. Water Science and Technology, 41 (1), 241-244.

CIW (1999). Individuele Behandeling van Afvalwater IBA-systemen (Discharge demands for Individual Wastewater Treatment Systems). Commissie Integraal Waterbeheer, Lelystad, The Netherlands.

Clarke J.M., Gillings M.R., Altavilla N., Beattie A.J. (2001). Potential problems with fluorescein diacetate assays of cell viability when testing natural producst for antimicrobial activity. Journal of Microbiological Methods, 46, 261-267.

Conkle J., White J.R., Metcalfe C.D. (2008). Reduction of pharmaceutically active compounds by a lagoon wetland wastewater treatment system in Southeast Louisiana. Chemosphere, 73, 1741-1748.

Cooper P. (1999). A review of the design and performance of vertical-flow and hybrid reed bed treatment systems. Water Science and Technology, 40 (3), 1-9.

Cooper P. (2005). The performance of vertical flow constructed wetland systems with special reference to the significance of oxygen transfer and hydraulic loading rates. Water Science and Technology, 51 (9), 81-90.

Cucarella V., Renman G. (2009). Phosphorus sorption capacity of filter materials used for on-site wastewater treatment determinated in batch experiments – A comparative study. Journal of Environmental Quality, 38, 381-392.

Cumbal L., Greenleaf J., Leun D., SenGupta A.K. (2003). Polymer supported inorganic nanoparticles: characterization and environmental applications. Reactive and Functional Polymers, 54, 167-180.

Decamp O., Warren A. (1998). Bacterivory in ciliates isolated from constructed wetlands (reed beds) used for wastewater treatment. Water Research, 32, 1989-1996.

Decamp O., Warren A., Sánchez R. (1999). The role of ciliated protozoa in subsurface flow wetlands and their potential as bioindicators. Water Science and Technology, 40 (3), 91-98.

Dignac M.F, Ginestet P., Rybacki D., Bruchet A., Urbain V., Scribe P. (2000). Fate of wastewater organic pollution during activated sludge treatment: nature of residual organic matter. Water Research, 34, 4185-4194.

Dong H., Qiang Z., Li T., Jin H., Chen W. (2012). Effect of artificial aeration on the performance of vertical-flow constructed wetland treating heavily polluted river water. Journal of Environmental Sciences, 24, 596-601.

Dunne E.J., Reddy R. (2005). Phosphorus biogeochemistry of wetlands in agricultural watersheds. In: Dunne, E.J., Reddy, R., Carton O.T. (Eds.), Nutrient management in agricultural watersheds: a wetland solution, Wageningen Academic Publishers, Wageningen, The Netherlands, pp. 105-119.

EEC (1991). Council Directive 91/271/EEC of 21 May 1991 concerning urban waste water treatment, Official Journal of the European Communities. L135, 30.5.1991, p. 40.

Eisenberg B., Nemcova E., Poblet R., Stokman A. (2014). Lima Ecological Infrastructure Strategy, Stuttgart, Germany. ISBN:978-3-00-047557-3.

Engloner A.I. (2009). Structure, growth dynamics and biomass of reed (*Phragmites australis*) – A review. Flora, 204, 331-346.

EPA (2009). Code of Practice: Wastewater treatment and disposal systems serving single households (p.e. less than or equal to 10). Environmental Protection Agency Ireland.

Fan J., Wang W., Zhang B., Guo Y., Ngo H.H., Guo W., Zhang J., Wu H. (2013a). Nitrogen removal in intermittently aerated vertical flow constructed wetlands: Impact of influent COD/N ratios. Bioresource Technology, 143, 461-466.

Fan J., Liang S., Zhang B., Zhang J. (2013b). Enhanced organics and nitrogen removal in batch-operated vertical flow constructed wetlands by combination of intermittent aeration and step feeding strategy. Environmental Science and Pollution Research, 20, 2448-2455.

Fan J., Zhang B., Zhang J., Ngo H.H., Guo W., Liu F., Guo Y., Wu H. (2013c). Intermittent aeration strategy to enhance organics and nitrogen removal in subsurface flow constructed wetlands. Bioresource Technology, 141, 117-122.

Fernández-Maldonado A.M. (2008). Expanding networks for the urban poor: Water and telecommunications services in Lima, Peru. Geoforum, 39, 1884-1896.

Finlay B.J., Rogerson A., Cowling A.J. (1988). A beginner's guide to the collection, isolation, cultivation and identification of freshwater protozoa. Culture Collection of Algae and Protozoa Publication, Freshwater Biological Association, Cumbria, UK, pp. 78.

FLL (2002). Guidelines for the Planning, Execution and Upkeep of Green-roof sites, 2002 edition, Forchungsgesellschaft Landschaftsentwicklung Landschaftsbau e.V. (FLL), Bonn.

Foladori P., Ortigara A.R.C., Ruaben J., Andreottola G. (2012). Influence of high organic loads during the summer period on the performance of hybrid constructed wetlands (VSSF + HSSF) treating domestic wastewater in the Alps region. Water Science and Technology, 65 (5), 890-897.

Foladori P., Ruaben J., Ortigara A.R.C. (2013). Recirculation or artificial aeration in vertical flow constructed wetlands: A comparative study for treating high load wastewater. Bioresource Technology, 149, 398-405.

Foladori P., Ruaben J., Ortigara A.R.C., Andreottola G. (2014). Batch feed and intermittent recirculation to increase removed loads in a vertical subsurface flow filter. Ecological Engineering, 70, 124-132.

Frazer-Williams R., Avery L., Winward G., Shirley-Smith C., Jefferson B. (2006). The Green Roof Water Recycling System - a novel constructed wetland for urban grey water recycling. In: Proceedings of 10th International conference on Wetland Systems for Water pollution Control, Lisbon, Portugal. 23 - 29th September 2006, Vol I, pp. 411-421.

Gallage C., Kodikara J., Uchimura T. (2013). Laboratory measurement of hydraulic conductivity functions of two unsaturated sandy soils during drying and wetting processes. Soils and Foundations, 53, 417-430.

García J., Aguirre P., Barragán J., Mujeriego R., Matamoros V., Bayona J.M. (2005). Effect of key design parameters on the efficiency of horizontal subsurface flow constructed wetlands. Ecological Engineering, 25, 405-418.

García J., Rousseau D.P.L., Morató J., Lesage E., Matamoros V., Bayona J.M. (2010). Contaminant removal processes in subsurface-flow constructed wetlands: a review. Critical Reviews in Environmental Science and Technology, 40, 561-661.

García J.A., Paredes D., Cubillos J.A. (2013). Effect of plants and the combination of wetland treatment type systems on pathogen removal in tropical climate conditions. Ecological Engineering, 58, 57-62.

Getter K.L., Rowe D.B. (2006). The role of extensive green roofs in sustainable development. HortScience, 41, 1276-1285.

Gerardi M.H. (2006). Wastewater bacteria. Wastewater microbiology series. Ed. John Wiley & Sons, Inc. New Jersey, USA, pp. 255.

Ghosh D., Gopal B. (2010). Effect of hydraulic retention time on the treatment of secondary effluent in a subsurface flow constructed wetland. Ecological Engineering, 36, 1044-1051.

Gómez Cerezo R., Suárez M.L., Vidal-Abarca M.R. (2001). The performance of a multi-stage system of constructed wetlands for urban wastewater treatment in a semiarid region of SE Spain. Ecological Engineering, 16, 501-517.

Gray N.F. (2004). Biology of wastewater treatment. Imperial College Press, London, UK, pp. 1421.

Green M., Friedler E., Ruskol Y., Safrai I. (1997). Investigation of alternative method for nitrification in constructed wetlands. Water Science and Technology, 35 (5), 63-70.

Green M., Friedler E., Safrai I. (1998). Enhancing nitrification in vertical flow constructed wetland utilizing a passive air pump. Water Research, 32, 3513-3520.

Green M., Shaul N., Beliavski M., Sabbah I., Ghattas B., Tarre S. (2006). Minimizing land requirement and evaporation in small wastewater treatmen systems. Ecological Engineering, 26, 266-271.

Gross A., Shmueli O., Ronen Z., Raveh E. (2007). Recycled vertical flow constructed wetland (RVFCW)—a novel method of recycling greywater for landscape irrigation in small communities and households. Chemosphere, 66, 916-923.

Gross A., Sklarz M.Y., Yakirevich A., Soares M.I.M. (2008). Small scale recirculating vertical flow constructed wetland (RVFCW) for the treatment and reuse of wastewater. Water Science and Technology, 58, 487-494.

Hahn M.W., Höfle M.G. (2001). Grazing of protozoa and its effect on populations of aquatic bacteria. FEMS Microbiology Ecology, 35, 113-121.

Halsey K., Arp D., Sayavedra L., Beedlow C. (2008). Nitrification Potential in Soils Workshop organized by Oregon State University, USA and Corvallis High School District 509J (http://nitrificationnetwork.org/potential.php).

Hammer D.A., Knight R.L. (1994). Designing constructed wetlands for nitrogen removal. Water Science and Technology, 29 (4), 15-27.

Haruta S., Takahashi T., Nishiguchi T. (1991). Basic studies on phosphorus removal by the contact aeration process using iron contactors. Water Science and Technology, 23 (4-6), 641-650.

Health Canada (2010). Canadian guidelines for domestic reclaimed water for use in toilet and urinal flushing, Health Canada, Ottawa, Ontario.

Her N., Amy G., McKnight D., Sohn J., Yoon Y. (2003). Characterization of DOM as a function of MW by fluorescence EEM and HPLC-SEC using UVA, DOC, and fluorescence detection. Water Research, 37, 4295-4303.

Hijosa-Valsero M., Matamoros V., Sidrach-Cardona R., Martín-Villacorta J., Bécares E., Bayona J.M. (2010). Comprehensive assessment of the design configuration of constructed wetlands for the removal of pharmaceuticals and personal care products from urban wastewaters. Water Research, 44, 3669-3678.

Hill T.H., Benson C.H. (1999). Hydraulic conductivity of compacted mine rock backfill. In: Proceedings of the 6th International Conference on Tailings and Mine Waste, Colorado, USA. 24 - 27th January 1999, pp. 373-380.

Hiscock K.M., Lloyd J.W., Lerner D.N. (1991). Review of natural and artificial denitrification of groundwater. Water Research, 25, 1099-1111.

Homar V., Ramis C., Romero R., Alonso S. (2010). Recent trends in temperature and precipitation over the Balearic Islands (Spain). Climatic Change, 98, 199-211.

Hu Y., Zhao Y., Zhao X., Kumar J.L.G. (2012). High rate nitrogen removal in an alum sludge-based intermittent aeration constructed wetland. Environmental Science and Technology, 46, 4583-4590.

Huws S.A., McBain A.J., Gilbert P. (2005). Protozoan grazing and its impact upon population dynamics in biofilm communities. Journal of Applied Microbiology, 98, 238-244.

Iasur-Kruh L., Hadar Y., Milstein D., Gasith A., Minz D. (2010). Microbial population and activity in wetland microcosms constructed for improving treated municipal wastewater. Microbial Ecology, 59, 700-709.

Imai A., Fukushima T., Matsushige K., Kim Y.H., Choi K. (2002). Characterization of dissolved organic matter in effluents from wastewater treatment plants. Water Research, 36, 859-870.

Ioris A.A.R. (2012a). The geography of multiple scarcities: Urban development and water problems in Lima, Peru. Geoforum, 43, 612-622.

Ioris A.A.R. (2012b). The neoliberalization of water in Lima, Peru. Political Geography, 31, 266-278.

Jamieson T.S., Stratton G.W., Gordon R., Madani A. (2003). The use of aeration to enhance ammonia nitrogen removal in constructed wetlands. Canadian Biosystems Engineering, 45, 1.9-1.14.

Jenkins D., Hermanowicz S.W. (1991). Principles of chemical phosphate removal, in: Sedlak, I.R. (Eds.), Phosphorus and nitrogen removal from municipal wastewater: principles and practice (2nd Edition), Lewis Publishers, CRC Press, LLC, Raton, FL, USA, pp. 91-110.

Ji G., Wang R., Zhi W., Liu X., Kong Y., Tan Y. (2012). Distribution patterns of denitrification functional genes and microbial floras in multimedia constructed wetlands. Ecological Engineering, 44, 179-188.

Kadlec R.H., Wallace S.D. (2009). Treatment Wetlands, 2nd ed. CRC Press, Taylor and Francis Group, Boca Raton, FL, USA, pp. 1016.

Kantawanichkul S., Neamkam P., Shutes R. (2001). Nitrogen removal in a combined system: vertical vegetated bed over horizontal flow sand bed. Water Science and Technology, 44 (11-12), 137-142.

Kantawanichkul S., Somprasert S., Aekasin U., Shutes R. (2003). Treatment of agricultural wastewater in two experimental combined constructed wetland systems in a tropical climate. Water Science and Technology, 48 (5), 199-205.

Kantawanichkul S., Somprasert S. (2005). Using a compact combined constructed wetland system to treat agricultural wastewater with high nitrogen. Water Science and Technology, 51 (9), 47-53.

Kivaisi A.K. (2001). The potential for constructed wetlands for wastewater treatment and reuse in developing countries: a review. Ecological Engineering, 16, 545-560.

Knowles P., Dotro G., Nivala J., García J. (2011). Clogging in subsurface-flow treatment wetlands: occurrence and contributing factors. Ecological Engineering, 37(2), 99-112.

Koroleff F. (1983). Simultaneous oxidation of nitrogen and phosphorus compounds by persulfate. Methods of Seawater Analysis. Verlag Chemie, Weinheim, Germany.

Kwon H.B., Lee C.W., Jun B.S., Yun J.D., Weon S.Y., Koopman B. (2004). Recycling waste oyster shells for eutrophication control. Resources, Conservation and Recycling, 41, 75-82.

Langergraber G., Harberl R., Laber J., Pressl A. (2003). Evaluation of substrate clogging processes in vertical flow constructed wetlands. Water Science and Technology, 48 (5), 25-34.

Lee C.W., Kwon H.B., Jeon H.P., Koopman B. (2009). A new recycling material for removing phosphorus from water. Journal of Cleaner Production, 17, 683-687.

Lee S.Y., Maniquiz M.C., Choi J.Y., Jeong S.M., Kim L.H. (2013). Seasonal nutrient uptake of plant biomass in a constructed wetland treating piggery wastewater effluent. Water Science and Technology, 67(6), 1317-1323.

Lüderitz V., Gerlach F. (2002). Phosphorus removal in different constructed wetlands. Acta Biotechnologica, 22, 91-99.

Mara D.D. (2006). Natural wastewater treatment. Manual of Best Practice. Chartered Institution of Water and Environmental Management (CIWEM), University of Leeds, London.

Matz C., Kjelleberg S. (2005). Off the hook - how bacteria survive protozoan grazing. Trends in Microbiology, 13, 302-307.

Mburu N., Tebitendwa S.M., van Bruggen J.J.A., Rousseau D.P.L., Lens P.N.L. (2013). Performance comparison and economics analysis of waste stabilization ponds and horizontal subsurface flow constructed wetlands treating domestic wastewater: A case study of the Juja sewage treatment works. Journal of Environmental Management, 128, 220-225.

Mburu N., Rousseau D.P.L., van Bruggen J.J.A. Thumbi G., Llorens E., García J., Lens P.N.L (2013). Reactive transport simulation in a tropical horizontal subsurface flow constructed wetland treating domestic wastewater. Science of the Total Environment, 449, 309-319.

MEM (2012). Documento promotor del subsector electricidad 2012, Ministerio de Energía y Minas, Peru (in Spanish).

Metcalf & Eddy (2003). Wastewater Engineering: Treatment and Reuse, 4th edn. McGraw Hill International Editions, Civil Engineering Series, Singapore, pp. 1819.

Molle P., Liénard A., Boutin C., Merlin G., Iwema A. (2005). How to treat raw sewage with constructed wetlands: an overview of the French systems. Water Science and Technology, 51 (9), 11-21.

Molle P. (2014). French vertical flow constructed wetlands: a need of a better understanding of the role of the deposit layer. Water Science and Technology, 69 (1), 106-112.

Monjeau C. (1901). Patent: Purifying water. United States US 681,884.

Moscoso Cavallini J.C. (2011). Estudio de opciones de tratamiento y reuso de aguas residuales en Lima metropolitana. Project report (www.lima-water.de/documents/jmoscoso_informe_.pdf) (in Spanish).

Nam S.N., Amy G. (2008). Differentiation of wastewater effluent organic matter (EfOM) from natural organic matter (NOM) using multiple analytical techniques. Water Science and Technology, 57 (7), 1009-1015.

NEN 6472 (1983). Water - Photometric determination of ammonium content. Dutch Normalization Institute. Delft, The Netherlands.

Nivala J., Hoos M.B., Cross C., Wallace S., Parkin G. (2007). Treatment of landfill leachate using an aerated, horizontal subsurface-flow constructed wetland. Science of the Total Environment, 380, 19-27.

Nivala J., Headley T., Wallace S., Bernhard K., Brix H., van Afferden M., Müller R.A. (2013). Comparative analysis of constructed wetlands: The design and construction of the ecotechnology research facility in Langenreichenbach, Germany. Ecological Engineering, 61, 527-543.

Nivala J., Wallace S., Headley T., Kassa K., Brix H., van Afferden M., Müller R. (2013). Oxygen transfer and consumption in subsurface flow treatment wetlands. Ecological Engineering, 61, 544-554.

Nur T., Johir M.A.H., Loganathan P., Nguyen T., Vigneswaran S., Kandasamy J. (2014). Phosphate removal from water using an iron oxide impregnated strong base anion exchange resin. Journal of Industrial and Engineering Chemistry, 20, 1301-1307.

Oberndorfer E., Lundholm J., Bass B., Coffman R.R., Doshi H., Dunnett N., Gaffin S., Köhler M., Liu K.K.Y., Rowe B. (2007). Green roofs as urban ecosystems: ecological structures, functions, and services. BioScience, 57 (10), 823-833.

O'Hogain S, McCarton L, Reid A., Turner J., Fox S. (2011). A review of zero discharge wastewater treatment systems using reed willow bed combinations in Ireland. Journal of Water Practice and Technology, 6 (3).

Ouellet-Plamondon C., Chazarenc F., Comeau Y., Brisson J. (2006). Artificial aeration to increase pollutant removal efficiency of constructed wetlands in cold climate. Ecological Engineering, 27, 258-264.

Pan B., Wu J., Pan B., Lv L., Zhang W., Xiao L., Wang X., Tao X., Zheng S. (2009). Development of plymer-based nanosized hydrated ferric oxides (HFOs) for enhanced phosphate removal from waste effluents. Water Research, 43, 4421-4429.

Park W.H., Polprasert C. (2008). Roles of oyster shells in an integrated constructed wetland system designed for P removal. Ecological Engineering, 34, 50-56.

Park N., Vanderford B.J., Snyder S.A., Sarp S., Kim S.D., Cho J. (2009). Effective controls of micropollutants included in wastewater effluent using constructed wetlands under anoxic conditions. Ecological Engineering, 35, 418-423.

Parmar N., Singh A., Ward O.P. (2001). Characterization of the combined effects of enzyme, pH and temperature treatments for removal of pathogens from sewage sludge. World Journal of Microbiology and Biotechnology, 17, 169-172.

Pérez Rubí M.A. (2014). Green wall for domestic wastewater treatment. UNESCO-IHE MSc. Thesis ES14.28.

Pintér P. (2014). Influence of recirculation in a pulse-fed Duplex constructed wetland used for domestic wastewater treatment. UNESCO-IHE MSc Thesis and International Master of Science in Environmental Technology and Engineering Erasmus Mundus Programme (Course N° 2011-0172).

Prigent S., Belbeze G., Paing J., Andres Y., Voisin J., Chazarenc F. (2013). Biological characterization and treatment performances of a compact vertical flow constructed wetland with the use of expanded schist. Ecological Engineering, 52, 12-18.

Prochaska C.A., Zouboulis A.I., Eskridge K.M. (2007). Performance of pilot-scale vertical-flow constructed wetlands, as affected by season, substrate, hydraulic load and frequency of application of simulated urban sewage. Ecological Engineering, 31, 57-66.

Puigagut J., Salvado H., García J. (2007). Effect of soluble and particulate organic compounds on microfauna the community in subsurface flow constructed wetlands. Ecological Engineering, 29, 280-286.

Rittmann B.E., Mayer B., Westerhoff P., Edwards M. (2011). Capturing the lost phosphorus. Chemosphere, 84, 846-853.

Robertson L.A., Kuenen J.G. (1984). Aerobic denitrification—old wine in new bottles?. Antonie van Leeuwenhoek Journal of Microbiology, 50, 525-544.

Rousseau D.P.L., Vanrolleghem P.A., De Pauw N. (2004). Model-based design of horizontal subsurface flow constructed treatment wetlands: a review. Water Research, 38, 1484-1493.

Saar R.A., Weber J.H. (1982). Fulvic acid: modifier of metal-anion chemistry. Environmental Science and Technology, 16, 510A-517A.

Sánchez-Monedero M.A., Mondini C., Cayuela M.L., Roig A., Contin M., De Nobili M. (2008). Fluorescein diacetate hydrolysis, respiration and microbial biomass in freshly amended soils. Biology and Fertility of Soils, 44, 885-890.

Seifert D., Engesgaard P. (2007). Use of tracer test to investigate changes in flow and transport properties due to bioclogging of porous media. Journal of Contaminant Hydrology, 93, 58-71.

Sengupta S., Pandit A. (2011). Selective removal of phosphorus from wastewater combined with its recovery as a solid-phase fertilizer. Water Research, 45, 3318-3330.

Shipin O., Koottatep T., Khanh N.T.T., Polprasert C. (2005). Integrated natural treatment systems for developing communities: low-tech N-removal through the fluctuating microbial pathways. Water Science and Technology, 51 (12), 299-306.

Sinclair J.L., Kampbell D.H., Cook M.L., Wilson J.T. (1993). Protozoa in subsurface sediments from sites contaminated with aviation gasoline or jet fuel. Applied and Environmental Microbiology, 59, 467-472.

Sklarz M.Y., Gross A., Yakirevich A., Soares M.I.M. (2009). A recirculating vertical flow constructed wetland for the treatment of domestic wastewater. Desalination, 246 (1-3), 617-624.

Sklarz M.Y., Gross A., Soares M.I.M., Yakirevich A. (2010). Mathematical model for analysis of recirculating vertical flow constructed wetlands. Water Research, 44, 2010-2020.

Shipin O., Koottatep T., Khanh N.T.T., Polprasert C. (2005). Integrated natural treatment systems for developing communities: low-tech N-removal through the fluctuating microbial pathways. Water Science and Technology, 51 (12), 299-306.

Sirivedhin T., Gray K.A. (2006). Factors affecting denitrification rates in experimental wetlands: field and laboratory studies. Ecological Engineering, 26, 167-181.

Sproul O.J. (1980). Critical review of virus removal by coagulation processes and pH modifications. U.S. Environmental Protection Agency, Cincinnati, Ohio.

Stewart P.S. (2003). Diffusion in biofilms. Journal of Bacteriology, 185, 1485-1491.

Streble H., Krauter D. (2006). Das Leben im Wassertropfen : Mikroflora und Mikrofauna des Süßwassers. Stuttgart: Franckh-Kosmos Verlags-GmbH & Co.

Sun G., Gray K.R., Biddlestone A.J., Allen S.J., Cooper D.J. (2003). Effect of effluent recirculation on the performance of a reed bed system treating agricultural wastewater. Process Biochemistry, 39, 351-357.

Sun G., Zhao Y., Allen S. (2005). Enhanced removal of organic matter and ammoniacal-nitrogen in a column experiment of tidal flow constructed wetland system. Journal of Biotechnology, 115, 189-197.

Sun G., Zhao Y., Allen S., Cooper D. (2006). Generating "Tide" in pilot-scale constructed wetlands to enhance agricultural wastewater treatment. Engineering in Life Sciences, 6, 560-565.

Svete L.E. (2012). Vegetated greywater treatment walls: design modifications for intermittent media filters. Norwegian University of Life Sciences MSc. thesis.

Tang X., Huang S., Scholz M., Li J. (2009). Nutrient removal in pilot-scale constructed wetlands treating eutrophic river water: assessment of plants, intermittent artificial aeration and polyhedron hollow polypropylene balls. Water, Air and Soil Pollution, 197, 61-73.

Tanner C.C., Kadlec R.H. (2003). Oxygen flux implications of observed nitrogen removal rates in subsurface-flow treatment wetlands. Water Science and Technology, 48 (5), 191-198.

Tanner C.C., Nguyen M.L., Sukias J.P.S. (2005). Nutrient removal by a constructed wetland treating subsurface drainage from grazed dairy pasture. Agriculture, Ecosystems and Environment, 105, 145-162.

Tawfik A., El-Gohary F., Ohashi A., Harada H. (2006). The influence of physical–chemical and biological factors on the removal of faecal coliform through down-flow hanging sponge (DHS) system treating UASB reactor effluent. Water Research, 40, 1877-1883.

Teemusk A., Mander Ü. (2007). Rainwater runoff quantity and quality performance from a greenroof: The effects of short-term events. Ecological Engineering, 30, 271-277.

Thon A., Kircher W., Thon I. (2010). Constructed wetlands on roofs as a module of sanitary environmental engineering to improve urban climate and benefit of the on site thermal effects. Miestų želdynų formavimas, 1, 191-196.

Thullner M., Mauclaire L., Schroth M.H., Kinzelbach W., Zeyer J. (2002). Interaction between water flow and spatial distribution of microbial growth in a two-dimensional flow field in saturated porous media. Journal of Contaminant Hydrology, 58, 169-189.

Transfer, The Steinbeis Magazine (2010). The magazine for Steinbeis Network employees and customers. Issue 2, p. 8. Edition 2/2010 ISSN 1864-1768.

Troesch S., Prost-Boucle S., Molle S., Leboeuf V., Esser D. (2010). Reducing the footprint of vertical flow constructed wetlands for raw sewage treatment: the Bi-filtre®solution. In: Proceedings of the 12th International IWA Conference on Wetland Systems for Water Pollution Control. Venice, Italy, October 4-8th, 2010. Vol. 2, pp. 1004-1010.

Troesch S., Salma F., Esser D. (2014). Constructed wetlands for the treatment of raw wastewater: the French experience. Water Practice and Technology, 9, 430-439.

UNI EN 12457-2 (2004). UNI Ente Nazionale Italiano di Unificazione, Legge 22 aprile 1941 N.633 e successivi aggiornamenti.

United Nations (2004). World Population to 2300. Department of Economic and Social Affairs, Population Division, United Nations, ST/ESA/SER.A/236, New York.

United Nations (2005). Population Challenges and Development Goals. Department of Economic and Social Affairs, Population Division, United Nations, ST/ESA/SER.A/248, New York.

UNICEF and WHO (2012). Progress on drinking water and sanitation: 2012 update. United Nations Children's Fund, World Health Organization, New York.

USEPA (2012). Guidelines for water reuse. EPA/600/R-12/618. U.S. Environmental Protection Agency, Washington D.C.

Vergara W., Deeb A., Leino I., Hansen M. (2010). Assessment of the impacts of climate change on mountain hydrology : development of a methodology through a case study in Peru. A World Bank study. Washington D.C.: The Worldbank. (http://documents.worldbank.org/curated/en/2010/09/17496676/assessment-impacts-climate-change-mountain-hydrology-development-methodology-through-case-study-peru)

Verhoeven J.T.A., Meuleman A.F.M. (1999). Wetlands for wastewater treatment: Opportunities and limitations. Ecological Engineering, 12, 5-12.

Vohla C., Koiv M., Bavor H.J., Chazarenc F., Mander Ü. (2011). Filter Materials for phosphorus removal from wastewater in treatment wetlands-A review. Ecological Engineering, 37, 70-89.

von Sperling M. (1996). Comparison among the most frequently used systems for wastewater treatment in developing countries. Water Science and Technology, 33 (1), 59-72.

Vymazal J. (2005). Horizontal sub-surface flow and hybrid constructed wetlands systems for wastewater treatment. Ecological Engineering, 25, 478-490.

Vymazal J. (2007). Removal of nutrients in various types of constructed wetlands. Science of the Total Environment, 380, 48-65.

Vymazal J., Kröpfelová L. (2008). Is concentration of dissolved oxygen a good indicator of processes in filtration beds of horizontal-flow constructed wetlands?. In: Wastewater treatment, plant dynamics and management in constructed and natural wetlands, Springer Netherlands, pp. 311-317.

Vymazal J., Kröpfelová L. (2009). Removal of organics in constructed wetlands with horizontal sub-surface flow: a review of the field experience. Science of the total environment, 407, 3911-3922.

Vymazal J. (2010). Constructed wetlands for wastewater treatment. Water, 2, 530-549.

Vymazal J. (2011). Constructed wetlands for wastewater treatment: Five decades of experience. Environmental Science and Technology, 45 (1), 61-69.

Vymazal J. (2013). The use of hybrid constructed wetlands for wastewater treatment with special attention to nitrogen removal: A review of a recent development. Water Research, 47, 4795-4811.

Walinga I., van Vark W., Houba V.J.G., van der Lee J.J. (1989). Soil and plant analysis. Part 7, Plant analysis procedures. Wageningen Agriculture University, Wageningen, USA, pp. 263.

Wallace S. (2001). Patent: System for removing pollutants from water. United States US 6,200,469 B1.

Wang Y., Zhou Q.H., Liang W., Xu D., Cai L.L., Tao M., Wu Z.B. (2010). Isolation and identification of a high-efficiency aerobic denitrifier and its denitrifying characteristic in constructed wetland. Journal of Agro-Environment Science, 29, 1193-1198 (in Chinese).

Wang Z., Dong J., Liu L., Zhu G., Liu C. (2013). Screening of phosphate-removing substrates for use in constructed wetlands treating swine wastewater. Ecological Engineering, 54, 54-65.

Weerakoon G.M.P.R., Jinadasa K.B.S.N., Herath G.B.B., Mowjood M.I.M., van Bruggen J.J.A. (2013). Impact of the hydraulic loading rate on pollutants removal in tropical horizontal subsurface flow constructed wetlands. Ecological Engineering, 61, 154-160.

Weerakoon G.M.P.R., Jinadasa K.B.S.N., Herath G.B.B., Mowjood M.I.M., van Bruggen J.J.A. (2013). Impact of the hydraulic loading rate on pollutants removal in tropical horizontal subsurface flow constructed wetlands. Ecological Engineering, 61, 154-160.

Wei L.L., Zhao Q.L., Xue S., Jia T., Tang F., You P.Y. (2009). Behavior and characteristics of DOM during a laboratory-scale horizontal subsurface flow wetland treatment: Effect of DOM derived from leaves and roots. Ecological Engineering, 35, 1405-1414.

WHO and UN-HABITAT (2010). Hidden cities: unmasking and overcoming health inequities in urban settings. World Health Organization, The WHO Centre for Health Development, Kobe, and United Nations Human Settlements Programme, Switzerland.

Winward G.P., Avery L.M., Frazer-Williams R., Pidou M., Jeffrey P. Stephenson T., Jefferson B. (2008). A study of the microbial quality of grey water and an evaluation of treatment technologies for reuse. Ecological Engineering, 32, 187-197.

Wong T.H. (2006). An Overview of Water Sensitive Urban Design Practices in Australia. Water Practice and Technology, 1 (01). doi: 10.2166/WPT.2006018.

Wong T.H.F. (2007). Water sensitive urban design—the journey thus far. Australian Journal of Water Resources, 10, 213-221.

Wu H., Zhang J., Wei R., Liang S., Li C., Xie H. (2013). Nitrogen transformations and balance in constructed wetlands for slightly river water treatment using different macrophytes. Environmental Science and Pollution Research, 20, 443-451.

Wu S., Kuschk P., Brix H., Vymazal J., Dong R. (2014). Development of constructed wetlands in performance intensifications for wastewater treatment: A nitrogen and organic matter targeted review. Water Research, 57, 40-55.

Wuertz S., Bishop P., Wilderer P. (2003). Biofilms in wastewater treatment: an interdisciplinary approach. IWA Publishing, pp 401.

Xu D., Li Y., Howard A., Guan Y. (2013). Effect of earthworm *Eisenia fetida* and wetland plants on nitrification and denitrification potentials in vertical flow constructed wetland. Chemosphere, 92, 201-206.

Yaghi N., Hartikainen H. (2013). Enhancement of phosphorus sorption onto light expanded clay aggregates by means of aluminum and iron oxide coatings. Chemosphere, 93, 1879-1886.

Ye F., Li Y. (2009). Enhancement of nitrogen removal in towery hybrid constructed wetland to treat domestic wastewater for small rural communities. Ecological Engineering, 35, 1043-1050.

Yoon G.L., Kim B.T., Kim B.O., Han S.H. (2003). Chemical-mechanical characteristics of crushed oyster-shell. Waste Management, 23, 825-834.

Zapater M., Gross A., Soares M.I.M. (2011). Capacity of an on-site recirculating vertical flow constructed wetland to withstand disturbances and highly variable influent quality. Ecological Engineering, 37, 1572-1577.

Zapater-Pereyra M., Dien van F., Bruggen van J.J.A., Lens P.N.L. (2013). Material selection for a constructed wetroof receiving pre-treated high strength domestic wastewater. Water Science and Technology, 68 (10), 2264-2270.

Zapater-Pereyra M., Gashugi E., Rousseau D.P.L., Alam M.R., Bayansan T., Lens P.N.L. (2014). Effect of aeration on pollutants removal, biofilm activity and protozoan abundance in conventional and hybrid horizontal subsurface-flow constructed wetlands. Environmental Technology, 35, 2086-2094.

Zapater-Pereyra M., Ilyas, H., Lavrnić S., Bruggen van J.J.A., Lens P.N.L. (2015a). Evaluation of the performance and the space requirement by three different hybrid constructed wetlands in a stack arrangement, Ecological Engineering, 82, 290-300.

Zapater-Pereyra M., Lavrnić S., Dien van F., Bruggen van J.J.A., Lens P.N.L. (2015b). Constructed wetroofs: a novel approach for the treatment and reuse of domestic wastewater at household level, Submitted to Journal of Environmental Management.

Zhai X., Piwpuan N., Arias C.A., Headley T., Brix H. (2013). Can root exudates from emergent wetland plants fuel denitrification in subsurface flow constructed wetland systems? Ecological Engineering, 61, 555-563.

Zhang L.Y., Zhang L., Liu Y.D., Shen Y.W., Liu H., Xiong Y. (2010). Effect of limited artificial aeration on constructed wetland treatment of domestic wastewater. Desalination, 250, 915-920.

Zhao Y.Q., Sun G., Allen S.J. (2004a). Purification capacity of a highly loaded laboratory scale tidal flow reed bed system with effluent recirculation. Science of the Total Environment, 330, 1-8.

Zhao Y.Q., Sun G., Lafferty C., Allen S.J. (2004b). Optimising performance of a novel reed bed system for the treatment of high strength agricultural wastewater. Water Science and Technology, 50 (8), 65-72.

Zhao Y.Q., Sun G., Allen S.J. (2004c). Anti-sized reed bed system for animal wastewater treatment: a comparative study. Water Research, 38, 2907-2917.

Zhao L., Zhu W., Tong Wei. (2009). Clogging processes caused by biofilm growth and organic particle accumulation in lab-scale vertical flow constructed wetlands. Journal of Environmental Science, 21, 750-757.

Zhao Y.Q., Babatunde A.O., Hu Y.S., Kumar J.L.G., Zhao X.H. (2011). Pilot field-scale demonstration of a novel alum sludge-based constructed wetland system for enhanced wastewater treatment. Process Biochemistry, 46, 278-283.

APPENDICES

APPENDIX A

Nanoparticle-beads batch experiment with neutral initial pH

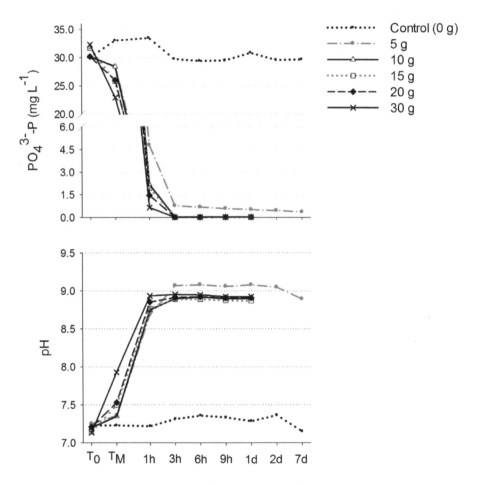

Figure A.1 Phosphorus removal (top) and pH (bottom) in the batch experiment conducted with different nanoparticle-beads weighs (Control (0 g), 5, 10, 15, 20 and 30 g). Experimental conditions: 450 mL of phosphate solution, 175 rpm, ambient temperature, pH manually fixed to neutral values at T_0, conducted in duplicates. Abbreviations: T_0, time zero (phosphorus source alone) and T_M, time immediately after addition of material.

APPENDIX B

Air flow in the aerated constructed wetlands

Required oxygen transfer rate (in g O_2 d^{-1}) for a complete aerobic treatment (Cooper, 2005; Kadlec and Wallace, 2009):

$$OTR_{required} = \text{flow rate} \times [4.6\,(NH_4\text{-}N)_{in} + (BOD)_{in}]$$

Eq. (B.1)

Oxygen transfer equation from Metcalf and Eddy (2003):

$$\frac{AOTR}{SOTR} = \left(\frac{\beta\, C_{S,T,H} - C_L}{C_{S,20}}\right) 1.024^{T-20}\; \alpha\; F$$

Eq. (B.2)

Where,

AOTR, actual oxygen transfer rate under field conditions (g O_2 d^{-1})
SOTR, standard oxygen transfer rate in tap water at 20°C and zero dissolved oxygen (g O_2 d^{-1})
β, correction factor for salinity and surface tension, typically 0.95 to 0.98
$C_{S,T,H}$, mean dissolved oxygen saturation concentration in clean water in aeration tank at temperature T and altitude H (mg L^{-1})
C_L, operating oxygen concentration (mg L^{-1})
$C_{S,20}$, dissolved oxygen saturation concentration in clean water at 20°C and 1 atm (mg L^{-1}). $C_{S,20}$ = 9.08 mg L^{-1}
1.024, temperature correction factor
T, operating temperature (°C)
α, oxygen transfer correction factor in wastewater vs. tap water, typically 0.3 to 1.2
F, fouling factor of air diffusers, typically 0.65 to 0.9

The air flow rate (L min^{-1}) is given by the following equation,

$$Q_{air} = \left(\frac{SOTR}{0.23 \times SOTE \times e_{air}}\right)$$

Eq. (B.3)

Where,

0.23, weight fraction of oxygen in air
SOTE, mean oxygen transfer efficiency in gravel-bed horizontal flow constructed wetland. This value is 4.7% per meter of water depth (Kadlec and Wallace, 2009)
e_{air}, air density at 0°C and 1 atm (mg L^{-1}). e_{air} = 1.2754 kg m^{-3}

As the air flow rate will be impacted by the site-elevation (H) and -temperature (T), the following equations should be used for corrections,

$$p = 101.33 \ (1 - 2.25577 \times 10^{-5} \ H)^{5.2558}$$

Eq. (B.4)

$$V_2 = V_1 \times \left(\frac{T_2}{T_1}\right) \text{ (from ideal gas law)}$$

Eq. (B.5)

Where,

p,   air pressure (kPa)
$V_{1,2}$,   volume at state 1 and 2 (L)
$T_{1,2}$,   temperature at state 1 and 2 (Kelvin)

The following steps are used to calculate the air flow rate necessary to be delivered by the air blower in the aerated system:

1. Use Equation B.1 to calculate the $OTR_{required}$. This $OTR_{required}$ is the amount of oxygen needed in the system to have a complete BOD_5 and NH_4^+-N removal (from the influent wastewater).

2. Because the OTR is affected by temperature, elevation and the wastewater characteristics, some corrections factors must be applied. Thus, use Equation B.2 to calculate the SOTR. The AOTR is obtained in the field, however, since the $OTR_{required}$ is the goal, the AOTR was replaced with the $OTR_{required}$. Values of β, C_L, α and F used were 0.95, 1 mg L^{-1}, 0.80 and 0.8, according to Scott Wallace (personal communication) and Boog et al. (2014). $C_{S,T,H}$, was obtained from the Appendix D of Metcalf and Eddy (2003) using T as 22°C (maximum room temperature) at 1 atm (760 mm of Hg).

3. Use Equation B.3 to convert the SOTR into air flow units (Q_{air}, L min^{-1}).

4. Use Equation B.4 to correct to site-elevation.

5. Use Equation B.5 to correct for site-temperature.

CHAPTER 4. Calculations from Section 4.2.2

Calculations were done for the second experimental period since the influent water had a higher organic loading rate than that in the first experimental period.

Knowing that (from Table 4.1),
Maximum influent COD = 340 mg L^{-1}
Maximum influent NH_4^+-N = 34 mg L^{-1}
Water flow = 16.3 L d^{-1}

Assuming a COD:BOD_5 ratio of 2, the influent BOD_5 would be 170 mg L^{-1}.

Step 1. Calculate $OTR_{required}$.

$OTR_{required} = 16.3 \times (4.6 \times 34 + 170)$
$OTR_{required} = 5320.32$ mg O_2 d^{-1}
$OTR_{required} = 5.32$ g O_2 d^{-1}

Step 2. Calculate SOTR. Replace AOTR with $OTR_{required}$.

$$\frac{AOTR}{SOTR} = \left(\frac{\beta\, C_{S,T,H} - C_L}{C_{S,20}}\right) 1.024^{T-20}\ \alpha\ F$$

$$\frac{AOTR}{SOTR} = \left(\frac{0.95 \times 8.73 - 1}{9.08}\right) 1.024^{22-20} \times 0.8 \times 0.8$$

$$\frac{AOTR}{SOTR} = 0.54$$

$$SOTR = \frac{5.32}{0.54}$$

$SOTR = 9.85$ g O_2 d^{-1}

Step 3. Convert to air flow rate units.

First, the water depth (in Section 4.2.1) was used to calculate SOTE for the aerated and hybrid system,
$SOTE_{Aerated} = 0.047 \times (0.38 - 0.04) = 0.01598$
$SOTE_{Hybrid} = 0.047 \times (0.20 - 0.04) = 0.00752$

Then values were replaced in Equation B.3

$$Q_{air-Aerated} = \left(\frac{9.85}{0.23 \times 0.01598 \times 1.2754}\right)$$
$Q_{air-Aerated} = 2.10$ m^3 d^{-1}
$Q_{air-Aerated} = 1.46$ L min^{-1}

$$Q_{air-Hybrid} = \left(\frac{9.85}{0.23 \times 0.00752 \times 1.2754}\right)$$
$Q_{air-Hybrid} = 4.47$ m^3 d^{-1}
$Q_{air-Hybrid} = 3.10$ L min^{-1}

Step 4. Correct for site-elevation. The elevation in The Netherlands is considered to be 0 m.a.s.l.
$p = 101.33 (1 - 2.25577 \times 10^{-5}\ H)^{5.2558}$
$P = 101.33$ kPa

$Q_{air-Aerated} = 1.46$ L min$^{-1} \times 100$ kPa $\div 101.33$ kPa $= 1.44$ L min^{-1}

$Q_{air\text{-}Hybrid} = 3.10 \text{ L min}^{-1} \times 100 \text{ kPa} \div 101.33 \text{ kPa} = 3.06 \text{ L min}^{-1}$

Step 5. Correct for site-temperature. Assumed air temperature of 18°C.

$$V_2 = V_{1\times} \left(\frac{18 + 273.15}{273.15} \right)$$

$$V_2 = V_{1\times} 1.066$$

$Q_{air\text{-}Aerated} = 1.44 \text{ L min}^{-1} \times 1.066 = 1.54 \text{ L min}^{-1}$
$Q_{air\text{-}Hybrid} = 3.06 \text{ L min}^{-1} \times 1.066 = \textbf{3.26 L min}^{-1}$

Finally, in theory, any air flow rate higher than 3.3 L min^{-1} would provide all the oxygen necessary to completely remove all BOD$_5$ and NH$_4^+$-N in both systems.

CHAPTER 5. Calculations from Section 5.2.1

Knowing that (from Fig. 5.2),
Maximum influent BOD$_5$ = 264 mg L^{-1}
Maximum influent NH$_4^+$-N = 68 mg L^{-1}
Water flow = 11.14 L d^{-1}

Step 1. Calculate OTR$_{required}$.

$OTR_{required} = 11.14 \times (4.6 \times 68 + 264)$
$OTR_{required} = 6425.6 \text{ mg O}_2 \text{ d}^{-1}$
$OTR_{required} = 6.43 \text{ g O}_2 \text{ d}^{-1}$

Step 2. Calculate SOTR.

$$\frac{AOTR}{SOTR} = \left(\frac{\beta\, C_{S,T,H} - C_L}{C_{S,20}} \right) 1.024^{T-20}\, \alpha\ F$$

$$\frac{AOTR}{SOTR} = \left(\frac{0.95 \times 8.73 - 1}{9.08} \right) 1.024^{22-20} \times 0.8 \times 0.8$$

$$\frac{AOTR}{SOTR} = 0.54$$

$$SOTR = \frac{6.43}{0.54}$$

$SOTR = 11.93 \text{ g O}_2 \text{ d}^{-1}$

Step 3. Convert to air flow rate units.

First, the water depth was used to calculate SOTE. The maximum water depth that the VF CW received was used.

$SOTE_{Aerated} = 0.047 \times 0.4 = 0.0188$

Then values were replaced in Equation B.3

$$Q_{air-Aerated} = \left(\frac{11.93}{0.23 \times 0.0188 \times 1.2754}\right)$$

$Q_{air-Aerated} = 2.16 \text{ m}^3 \text{ d}^{-1}$

$Q_{air-Aerated} = 1.50 \text{ L min}^{-1}$

Step 4. Correct for site-elevation. The elevation in The Netherlands is considered to be 0 m.a.s.l.

$p = 101.33 (1 - 2.25577 \times 10^{-5} \text{ H})^{5.2558}$

$P = 101.33 \text{ kPa}$

$Q_{air-Aerated} = 1.50 \text{ L min}^{-1} \times 100 \text{ kPa} \div 101.33 \text{ kPa} = 1.48 \text{ L min}^{-1}$

Step 5. Correct for site-temperature. Assumed air temperature of 18°C.

$$V_2 = V_{1 \times} \left(\frac{18 + 273.15}{273.15}\right)$$

$V_2 = V_{1 \times} 1.066$

$Q_{air-Aerated} = 1.48 \text{ L min}^{-1} \times 1.066 = \textbf{1.58 L min}^{-1}$

Finally, in theory, any air flow rate higher than 1.6 L min^{-1} would provide all the oxygen necessary to completely remove all BOD$_5$ and NH$_4^+$-N in the Aerated Duplex-CW.

CHAPTER 6. Calculations from Section 6.2.2

Calculations were done for the higher domestic wastewater strength (WW^{++})

Knowing that (from Figs. 6.3 and 6.4),

Influent COD = 796 mg L^{-1}

Influent NH$_4^+$-N = 54 mg L^{-1}

Water flow = 11.14 L d^{-1}

Assuming a COD:BOD$_5$ ratio of 2, the influent BOD$_5$ would be 398 mg L^{-1}.

Step 1. Calculate OTR$_{required}$.

$OTR_{required} = 11.14 \times (4.6 \times 54 + 398)$

$OTR_{required} = 7204.57 \text{ mg O}_2 \text{ d}^{-1}$

$OTR_{required} = 7.20 \text{ g O}_2 \text{ d}^{-1}$

Step 2. Calculate SOTR.

$$\frac{AOTR}{SOTR} = \left(\frac{\beta \, C_{S,T,H} - C_L}{C_{S,20}}\right) 1.024^{T-20} \, \alpha \, F$$

$$\frac{AOTR}{SOTR} = \left(\frac{0.95 \times 8.73 - 1}{9.08}\right) 1.024^{22-20} \times 0.8 \times 0.8$$

$$\frac{AOTR}{SOTR} = 0.54$$

$$SOTR = \frac{7.20}{0.54}$$

$SOTR = 13.37$ g O_2 d^{-1}

Step 3. Convert to air flow rate units.

First, the water depth was used to calculate SOTE. The maximum water depth that the VF CW received was used.

$SOTE_{Aerated} = 0.047 \times 0.4 = 0.0188$

Then values were replaced in Equation B.3

$$Q_{air-Aerated} = \left(\frac{13.37}{0.23 \times 0.0188 \times 1.2754}\right)$$
$Q_{air-Aerated} = 2.42$ m^3 d^{-1}
$Q_{air-Aerated} = 1.68$ L min^{-1}

Step 4. Correct for site-elevation. The elevation in The Netherlands is considered to be 0 m.a.s.l.
$p = 101.33 (1 - 2.25577 \times 10^{-5}$ H$)^{5.2558}$
$P = 101.33$ kPa

$Q_{air-Aerated} = 1.68$ L min^{-1} $\times$ 100 kPa $\div$ 101.33 kPa $= 1.66$ L min^{-1}

Step 5. Correct for site-temperature. Assumed maximum air temperature (in greenhouse) of 40°C.

$$V_2 = V_{1 \times} \left(\frac{40 + 273.15}{273.15}\right)$$

$V_2 = V_{1 \times} 1.15$

$Q_{air-Aerated} = 1.66$ L min^{-1} $\times$ 1.15 = **1.90 L min^{-1}**

Finally, in theory, any air flow rate higher than 1.9 L min^{-1} would provide all the oxygen necessary to completely remove all BOD$_5$ and NH$_4^+$-N to all Aerated Duplex-CW treating WW^{++}.

APPENDIX C

Constructed wetlands footprint calculations based in first order equations

First order equations from Kadlec and Wallace (2009):

$$q = \frac{-k_T}{\ln\left(\frac{C_e - C^*}{C_i - C^*}\right)}$$

Eq. (C.1)

$$A = \left(\frac{-Q}{k_T}\right) \times \ln\left(\frac{C_e - C^*}{C_i - C^*}\right)$$

Eq. (C.2)

Where,

q, hydraulic loading rate (m d^{-1})
A, area (m^2)
Q, water flow (m^3 d^{-1})
k_T, first order rate constant at temperature T (m d^{-1})
C_i, initial concentration (mg L^{-1})
C_e, effluent concentration (mg L^{-1})
C^*, background concentration (mg L^{-1})

The following steps are used to calculate the footprint ratio between the conventional (Control) and "non-conventional" (e.g. Aerated, Recirculating and Hybrid/Duplex-CW) constructed wetlands, in order to find out if the area was reduced due to the add-ons:

1. Use Equation C.1 to calculate k_T of each constructed wetland. Values of q, C_e and C_i are known. Assume a value for C^*. In this thesis we assume $C^* = 0$ mg L^{-1}. Do this for all relevant parameters (e.g. COD, TSS, NH_4^+-N, TN and TP/PO_4^{3-}-P).

2. To compare the area needed by the non-conventional systems to that needed by the Control system, assume that the non-conventional system achieved the C_e of that in the Control system for a particular parameter. Use Equation C.2 to calculate A. Values of Q, k_T, C_e (of the Control) and C_i are known. Assume a value for C^*. In this thesis we assume $C^* = 0$ mg L^{-1}.

3. Compare the values obtained in Step 2 vs. the area needed by the Control system (design value) by calculating the ratio $A_{control}$:$A_{non-conventional}$. In this thesis the design area of the Control system ($A_{control}$) is 0.24 m^2. The calculated value (ratio) indicates how smaller the non-conventional system could have been (to reach the control effluent quality) as compared to the control system. If the ratio equals 1, the non-conventional system needs the same area as the conventional (control) system.

CHAPTER 4. Calculations from Section 4.4.3

Knowing that,

$A = 0.24$ m^2
$Q = 0.0163$ m^3 d^{-1}
$q = 0.068$ m d^{-1}
Concentration values of COD, TSS, NH$_4^+$-N, TN and TP were obtained from Table 4.1.

Step 1. First order rate constant (k_T, in m d^{-1}) for the three tested horizontal flow constructed wetlands.

Parameter	Control	Aerated	Hybrid
First experimental period - 10.5 g COD m^{-2}d^{-1}			
COD	0.081	0.116	0.087
TSS	0.120	0.199	0.175
NH$_4^+$-N	0.017	0.225	0.109
TN	0.015	-0.003	0.007
PO$_4^{3-}$-P	-0.002	0.001	-0.0004
Second experimental period - 19.7 g COD m^{-2}d^{-1}			
COD	0.079	0.152	0.118
TSS	0.151	0.198	0.166
NH$_4^+$-N	0.006	0.301	0.057
TN	0.018	0.028	0.035
PO$_4^{3-}$-P	-0.008	-0.011	-0.009

A negative value indicates that $C_e > C_i$.

Step 2. Area (m^2) needed by the aerated and hybrid horizontal flow constructed wetlands to reach the effluent quality achieved by the control system.

Parameter	Control*	Aerated	Hybrid
First experimental period - 10.5 g COD m^{-2}d^{-1}			
COD	0.24	0.167	0.223
TSS	0.24	0.145	0.165
NH$_4^+$-N	0.24	0.019	0.038
TN	0.24	-1.198	0.532
PO$_4^{3-}$-P	0.24	-0.428	1.302
Second experimental period - 19.7 g COD m^{-2}d^{-1}			
COD	0.24	0.126	0.162
TSS	0.24	0.183	0.219
NH$_4^+$-N	0.24	0.005	0.026
TN	0.24	0.154	0.122
PO$_4^{3-}$-P	0.24	0.170	0.203

*Actual area of the experimental setup. It was not calculated.

Step 3. Ratio $A_{control}$:$A_{aerated/hybrid}$.

Parameter	Aerated	Hybrid
First experimental period - 10.5 g COD m^{-2}d^{-1}		
COD	**1.4**	1.1
TSS	1.7	1.5
NH$_4^+$-N	**12.9**	6.3
TN	-0.2	0.5
PO$_4^{3-}$-P	-0.6	0.2
Second experimental period - 19.7 g COD m^{-2}d^{-1}		
COD	**1.9**	1.5
TSS	1.3	1.1
NH$_4^+$-N	**48.7**	9.2
TN	1.6	2.0
PO$_4^{3-}$-P	1.4	1.2

Bold numbers are values mentioned in Chapter 4.

From that we can say, for example, that the aerated constructed wetland could have been 1.4 times smaller than the control system to reach the same effluent COD concentration.

CHAPTER 5. Calculations from Section 5.4.1

Knowing that,
A = 0.24 m^2
Q = 0.01114 m^3 d^{-1}
q = 0.046 m d^{-1}
Concentration values of COD, TSS, NH$_4$$^+$-N, TN and TP were obtained from Figure 5.2.

Step 1. First order rate constant (k_T, in m d^{-1}).

Parameter	Control	Aerated	Recirculating
COD	0.083	0.081	0.099
BOD$_5$	0.084	0.073	0.081
TSS	0.048	0.057	0.063
NH$_4$$^+$-N	0.124	0.079	0.124
TN	0.096	0.067	0.025

Step 2. Area (m^2) needed by the Aerated and Recirculating Duplex-CW to reach the effluent quality achieved by the Control system.

Parameter	Control**	Aerated	Recirculating
COD	0.24	0.25	0.20
BOD$_5$	0.24	0.28	0.25
TSS	0.24	0.20	0.18
NH$_4$$^+$-N	0.24	0.38	0.24
TN	0.24	0.34	0.91

**Actual area of the experimental setup. It was not calculated.

Step 3. Ratio $A_{Control}:A_{Aerated/Recirculating}$.

Parameter	Aerated	Recirculating
COD	1.0	1.2
BOD$_5$	0.9	1.0
TSS	1.2	1.3
NH$_4$$^+$-N	0.6	1.0
TN	0.7	0.3

From that we can say, for example, that the Aerated system could have been 1.2 times smaller than the Control system to reach the same effluent TSS concentration. In general, no area reduction is achieved by the use of the intensified systems.

CHAPTER 6. Calculations from Section 6.4.6 done only for the Fill and Drain configuration

Knowing that,
A = 0.24 m^2
Q = 0.01114 m^3 d^{-1}
q = 0.046 m d^{-1}
Concentration values of COD, TSS, NH$_4$$^+$-N, TN and TP were obtained from Figures 6.3 and 6.4.

Step 1. First order rate constant (k_T, in m d^{-1}).

Parameter	WW		WW$^+$		WW^{++}		WW$^{++}_A$	
	VF CW (Control)	Duplex-CW	VF CW (Control)	Duplex-CW	VF CW (Control)	Duplex-CW	VF CW (Control)	Duplex-CW
COD	0.084	0.095	0.102	0.124	0.099	0.114	0.103	0.135
TSS	0.096	0.111	0.083	0.124	0.076	0.084	0.081	0.102
NH$_4^+$-N	0.087	0.088	0.046	0.060	0.032	0.037	0.051	0.059
TN	0.028	0.060	0.028	0.079	0.024	0.070	0.021	0.058
TP	0.040	0.074	0.038	0.044	0.034	0.027	0.046	0.050

Step 2. Area (m^2) needed by the Duplex-CW to reach the effluent quality achieved by the VF CW (Control).

Parameter	All WW periods VF CW** (Control)	WW Duplex-CW	WW$^+$ Duplex-CW	WW^{++} Duplex-CW	WW$^{++}_A$ Duplex-CW
COD	0.24	0.211	0.197	0.209	0.183
TSS	0.24	0.207	0.161	0.217	0.192
NH$_4^+$-N	0.24	0.236	0.182	0.208	0.207
TN	0.24	0.112	0.085	0.082	0.086
TP	0.24	0.131	0.206	0.308	0.222

**Actual area of the experimental setup. It was not calculated.

Step 3. Ratio $A_{VF\ CW(Control)}:A_{Duplex-CW}$.

Parameter	WW Duplex-CW	WW$^+$ Duplex-CW	WW^{++} Duplex-CW	WW$^{++}_A$ Duplex-CW
COD	1.1	1.2	1.1	1.3
TSS	1.2	1.5	1.1	1.3
NH$_4^+$-N	1.0	1.3	1.2	1.2
TN	**2.1**	**2.8**	**2.9**	**2.8**
TP	1.8	1.2	0.8	1.1

Bold numbers are values mentioned in Chapter 6.

From that we can say, for example, that the Fill and Drain Duplex-CW could have been 2.1 times smaller than the VF CW alone to reach the same effluent TN concentration.

APPENDIX D

Carbon source as electron donor for denitrification

D.1. Batch experiment to determine the type of water to be used as a carbon source to enhance denitrification in the Recirculating HFF

The potential denitrification (PDR) test was employed to select the type of water that could be used as a carbon source to enhance denitrification. The method is explained in Section 5.2.3.2. The carbon sources options were: influent wastewater, filtered (with a piece of cloth) influent wastewater, VF CW effluent and HFF effluent. Each of them was used to prepare the incubation solution needed, replacing the carbon source (glucose) explained in the methodology. The rest was kept identical. Four flasks containing the same sand sample were mixed with each of the prepared incubation solutions, capped, flushed with nitrogen gas and incubated for 48 h (Section 5.2.3.2). Two extra flasks (control) were prepared identically but using the glucose incubation solution (0.02 M and 0.005 M). This experiment was conducted once, in duplicate. Additionally, the initial COD from all incubation solutions was determined. The VF CW effluent, HFF effluent and sand used in this experiment were obtained from the Control HFF.

Results showed that the higher the COD concentration (and C/N ratio) of glucose, the higher the PDR (Fig. D.1). Glucose is a simple organic compound that microorganisms can readily metabolise (Gerardi, 2006), thus it is used as a positive control in the PDR methodology (Fig. D.1). The PDR of the other carbon sources varied from 0 (effluent) to 3.5 mg kg^{-1} h^{-1} (**filtered influent**) (Fig. D.1), despite their COD initial concentration and C/N ratio were similar. Hence, the **filtered influent** was selected as the type of water to be used as a carbon source to enhance denitrification in the Recirculating HFF.

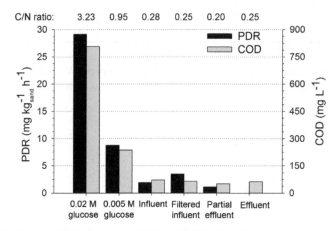

Figure D.1 Potential denitrification rate (PDR) and initial chemical oxygen demand (COD) in the batch experiment. The C/N ratio was calculated using COD and NO$_3^-$-N concentrations.

D.2 Calculations of volume of water as a carbon source to enhance denitrification in the Recirculating HFF

The following equation, given by Gray (2004) represents the amount of carbon required for denitrification,

Eq. (D.1)

C = 2.47 No + 1.52 Ni + 0.87 DO

Where,
C, carbon required for denitrification [mg L^{-1}]
No, nitrate concentration [mg L^{-1}]
Ni, nitrite concentration [mg L^{-1}]
DO, dissolved oxygen concentration [mg L^{-1}]

In Equation D.1, the value 2.47 indicates the methanol to nitrate ratio. Thus, methanol is considered the electron acceptor for denitrification. However, in wastewater, organic matter is the carbon source for denitrification. In that case, the following equation (for denitrification) given by Metcalf and Eddy (2003) is needed,

$$C_{10}H_{19}O_3N + 10NO_3^- \rightarrow 5N_2 + 10\ CO_2 + H_2O + 3NH_3 + 10\ OH^- \qquad \text{Eq. (D.2)}$$

The term $C_{10}H_{19}O_3N$ represents the biodegradable organic matter in wastewater. Calculating the organic matter to nitrate ratio from Equation D.2, it is obtained,

1 $C_{10}H_{19}O_3N$ / 10 NO_3^- = 0.1

Hence, Equation D.1 can be modified in order to consider organic matter as the electron acceptor for denitrification,

C = 0.1 No + 1.52 Ni + 0.87 DO   Eq. (D.3)

Now, it is necessary to replace No, Ni and DO in Equation D.3 to obtain C. The values are displayed in Table D.1. The carbon concentration required to denitrify the nitrate concentrations not removed by the HFF is: $C_{control}$ = 2.71 mg L^{-1} and $C_{recirculating}$ = 5.49 mg L^{-1} (Table D.1).

Table D.1 Mean nitrate, nitrite and dissolved oxygen concentrations (in mg L^{-1}) in the effluent of the Control and Recirculating systems

Parameter	Control effluent	Recirculating effluent
Nitrate[*]	6.4	31.0
Nitrite[*]	0.10	0.07
Dissolved oxygen[*]	2.20	2.63
Carbon (from Eq. D.3)	**2.71**	**5.49**

[*]Nitrate and dissolved oxygen concentrations were taken from Figure 5.2. Nitrite concentration was from measurements conducted, not displayed elsewhere.

The following schematic representation (Fig. D.2) is useful to understand how to calculate the volume of the filtered influent (Section D.1) used a carbon source to enhance denitrification in the HFF. The total volume entering the HFF was kept constant (39 L).

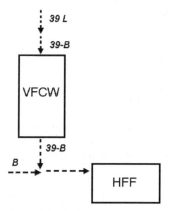

Figure D.2 Schematic representation of the influent volume (39-B) and filtered influent volume (B) flowing to the vertical flow constructed wetland (VF CW) and horizontal flow filter (HFF).

The mass of carbon needed by the volume 39-B should be identical to the mass of carbon provided by the filtered influent (carbon source, Section D.1) (B). From analysis, it is known that the influent wastewater has a dissolved organic carbon of ~42 mg L^{-1}, thus:

For the Control system,
42 mg L^{-1} × B = 2.71 mg L^{-1} × (39-B)
B = 2.4 L

For the Recirculating system,
42 mg L^{-1} × B = 5.49 mg L^{-1} × (39-B)
B = 4.5 L

Finally, the applied bypass volume was in the range of 2.4-4.5 L (average 3.5 L).

APPENDIX E

Duplex-Constructed Wetland

Chapter 5

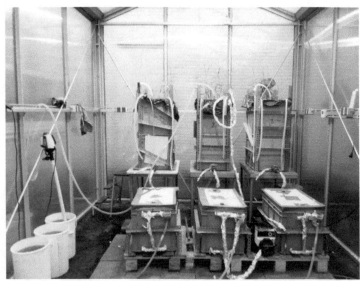

Left: Fill and drain, Middle: Stagnant batch, Right: Free drain
Horizontal flow filter in front of the vertical flow constructed wetlands for sampling purposes.
Photo credit: Violet Namakula

Air bubbles distribution experiment in the three tested Duplex-CWs

Aerated   Control   Recirculating

Chapter 6

Stagnant batch: elbow

Aeration pipes

Left: Fill and drain, Middle: Stagnant batch, Right: Free drain

Constructed wetroof

2012

09 May (Construction finalized)

Photo credit: Frank van Dien

04 July

07 June

2013

23 April

14 May

11 June

20 August

04 September

10 December

2014
23 April

Photo credit: Frank van Dien

APPENDIX F

Comparison of the potential nitrification rate using different NH_4^+-N concentrations in the incubation solution

Three different incubation solutions were tested:

- *Solution 1*: 0.2 M KH_2PO_4, 0.2 M K_2HPO_4 and 0.007 M $(NH_4)_2SO_4$ (200 mg NH_4^+-N L^{-1}) were mixed in a ratio of 3:37:30. The solution had a final concentration of approximately 150 mg NH_4^+-N L^{-1}. Adapted from Xu et al. (2013).
- *Solution 2*: 0.2 M KH_2PO_4, 0.2 M K_2HPO_4 and 0.014 M $(NH_4)_2SO_4$ (400 mg NH_4^+-N L^{-1}) were mixed in a ratio of 3:37:30. The solution had a final concentration of approximately 300 mg NH_4^+-N L^{-1}. Adapted from Xu et al. (2013).
- *Solution 3*: 0.0018 M KH_2PO_4 and 0.014 M K_2HPO_4 prepared together in 1 L volumetric flask. Separately, 0.063 M $(NH_4)_2SO_4$ (1764 mg NH_4^+-N L^{-1}) was prepared. The two solutions were mixed in a ratio 150:1. The solution had a final concentration of approximately 12 mg NH_4^+-N L^{-1}. Adapted from Halsey et al. (2008).

The following procedure was done for each solution: four glass bottles were filled with 120 mL of a solution and approximately 10 g of sand (n = 4). Two extra bottles filled only with 120 mL solution were used as the Solution-Control (n = 2). A total of 18 bottles (6 per solution).

Four extra bottles were filled with 10 g of sand and 120 mL of distilled water, to check if the sand is leaching nitrate to the incubation solution (Control) (n = 4).

All bottles were placed in the rotary shaker at 175 rpm in a 30°C room. NO_3^--N samples were taken at time 0 h and 48 h. The potential nitrification rate (PNR) was calculated as Xu et al. (2013).

Results revealed that Solution 1 and 2, despite the different NH_4^+-N concentrations, gave similar results. Solution 3, with an initial concentration of ~12 mg NH_4^+-N L^{-1} showed the highest value (Fig. F.1). The Control, did not show any nitrogen leachate (Fig. F.1) and the Solution-Control was stable during the 48 h of incubation (data not shown).

This preliminary experiment demonstrated that the PNR result varied depending on the initial NH_4^+-N concentration used. Since Solution 1 and 2 gave similar values, selecting any of them was appropriate. Solution 2 was used in this thesis for PNR analysis.

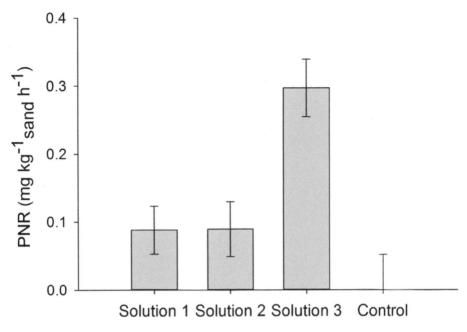

Figure F.1 Potential nitrification rate (PNR) of the different incubation solutions tested.

ACRONYMS

ADR	Actual denitrification rate
ANOVA	Analysis of variance
ANR	Actual nitrification rate
BOD$_5$	5-day biochemical oxygen demand
CC	Crushed coral
COD	Chemical oxygen demand
CW	Constructed wetland
CWR	Constructed wetroof
DO	Dissolved oxygen
DOC	Dissolved organic carbon
EC	Electrical conductivity
Em	Emission
EPS	Expanded polystyrene
Ex	Excitation
FDA	Fluorescein diacetate
Fill&D	Fill and drain
FreeD	Free drain
FWS	Free-water surface
GR	Green roof
GROW	Green roof water recycling system
HAIX	Hybrid anion exchanger
HF	Horizontal flow
HFF	Horizontal flow filter
HFO	Hydrated ferric oxide
HLR	Hydraulic loading rate
HRT	Hydraulic retention time
LBC	Load bearing capacity
LECA	Light expanded clay aggregate
MS	Mussel-shells
NB	Nanoparticle-beads
NWTS	Natural wastewater treatment system
OCR	Oxygen consumption rate

OLR	Organic loading rate
OS	Oyster-shells
PDR	Potential denitrification rate
PE	Population equivalent
PNR	Potential nitrification rate
PLA	Polylactic acid
SSF	Sub-surface water flow
StagB	Stagnant batch
TN	Total nitrogen
TP	Total phosphorus
TSS	Total suspended solids
VF	Vertical flow
WSUD	Water sensitive urban design
WW	Primary settled domestic wastewater
$WW^{+,++}$	Primary settled domestic wastewater with peptone
WW^{++}_A	Primary settled domestic wastewater with peptone and aeration

Publications, conferences and awards

Peer-reviewed journals

Zapater-Pereyra M., Kyomukama E., Namakula V., Bruggen van J.J.A., Lens P.N.L., "The effect of aeration and recirculation on a sand based hybrid constructed wetland treating low strength domestic wastewater", Submitted to Environmental Technology.

Zapater-Pereyra M., Lavrnić S., Dien van F., Bruggen van J.J.A., Lens P.N.L., "Constructed wetroofs: a novel approach for the treatment and reuse of domestic wastewater at household level", Submitted to Journal of Environmental Management.

Zapater-Pereyra M., Ilyas, H., Lavrnić S., Bruggen van J.J.A., Lens P.N.L. (2015), "Evaluation of the performance and the space requirement by three different hybrid constructed wetlands in a stack arrangement", Ecological Engineering, 82, 290-300.

Zapater-Pereyra M., Malloci E., Bruggen van J.J.A., Lens P.N.L. (2014), "Use of marine and engineered materials for the removal of phosphorus from secondary effluent", Ecological Engineering, 73, 635-642.

Zapater-Pereyra M., Gashugi E., Rousseau D.P.L., Alam M.R., Bayansan T., Lens P.N.L. (2014), "Effect of aeration on pollutants removal, biofilm activity and protozoan abundance in conventional and hybrid horizontal subsurface-flow constructed wetlands", Environmental Technology, 35, 2086-2094.

Zapater-Pereyra M., Dien van F., Bruggen van J.J.A., Lens P.N.L. (2013), "Material selection for a constructed wetroof receiving pre-treated high strength domestic wastewater", Water Science and Technology, 68 (10), 2264-2270.

Conference proceedings and oral presentations

Lavrnić S., *Zapater-Pereyra M.*, Bruggen van J.J.A., Dien van F., Lens P.N.L. (2014), "Nutrient flow and hydrology of a 9-cm-deep constructed wetroof", Proceedings of the 14th International Conference on Wetland Systems for Water Pollution Control, Shanghai, China, 852-862 [oral presentation by Lavrnić S.].

Zapater-Pereyra M., Namakula V., Kyomukama E., Bruggen van J.J.A., Lens P.N.L. (2014), "Nitrification and denitrificatino in a Duplex-constructed wetland", 14th International Conference on Wetland Systems for Water Pollution Control, Shanghai, China [oral presentation].

Zapater-Pereyra M., Ilyas H., Bruggen van J.J.A., Lens P.N.L. (2013), "Reducing the constructed wetland (CW) footprint by means of a Duplex-CW". Proceedings of the 5th International Symposium on Wetland Pollutant Dynamics and Control, WETPOL, Nantes, France. Book of abstracts, O.148 [oral presentation].

Zapater-Pereyra M., Lavrnić S, Bruggen van J.J.A., Dien van F, Lens P.N.L. (2013), "Can a 9-cm-depth horizontal flow constructed wetland, or "constructed wetroof", treat domestic wastewater?". BENELUX Young Water Professional 3rd Regional Conference, Luxembourg, Oct 2013 [oral presentation].

Zapater-Pereyra M., Malloci E., Bruggen van J.J.A., Lens P.N.L. (2012), "Engineered and marine materials, potential phosphorus removal options for constructed wetlands", 13th International Conference on Wetland Systems for Water Pollution Control, Perth, Australia [oral presentation].

Zapater-Pereyra M., Dien van F., Bruggen van J.J.A., Lens P.N.L. (2012), "Material selection for a constructed wetroof treating wastewater", Proceedings of the 13th International Conference on Wetland Systems for Water Pollution Control, Perth, Australia [oral presentation by Dien van F.].

Rousseau D.P.L., Gashugi E., Bayansan T. Alam R., *Zapater M.*, Lens P. (2011), "A comparison of organic matter and nutrients removal in HSSF CW under different aeration regimes", Proceedings of the Joint meeting of Society of Wetland Scientists, WETPOL and Wetland Biogeochemistry Symposium". Book of Abstracts, p. 278 [oral presentation by Rousseau D.P.L.].

Awards

Green Talents Award, International Forum for High Potentials in Sustainable Development. An initiative of the Federal Ministry of Education and Research and FONA – Research for Sustainable Development, Germany, 2011.

Biography

Maribel Zapater Pereyra was born in Piura, Peru on the 26th January 1984. She graduated from the School of Civil Engineering in 2005 from the University of Piura, Piura, Peru. Afterwards, she moved to Israel for her MSc. studies. In 2009 she obtained her MSc in Desert Studies from the University of the Negev, Sde Boqer, Israel. During that time, she learnt about natural wastewater treatment systems, got fascinated with the topic and decided to work as a Research Assistant in a project between Israel and Peru investigating different aspects of constructed wetlands.

In 2010 she initiated her PhD research in The Netherlands with the aim of designing and developing novel constructed wetlands with low space requirement. During her PhD studies she participated in different international conferences and supervised 6 MSc. students. She was also involved in internships with the company ECOFYT (The Netherlands) and with the University of Stuttgart (Germany). The first dealing with the application and construction of full scale constructed wetlands in The Netherlands and the second, with the combination of wastewater treatment systems within the landscape of the city of Lima (Peru).

Ms. Zapater Pereyra has 8 years of laboratory experience, a large experience in international conferences, an expanded international network and several scientific publications including proceedings, peer-reviewed journals and a collaboration in a book chapter. She also won in 2011 the recognized international award given by the German Federal Ministry of Education and Research, called Green Talents. She continues to be an active young professional with an interest in the applied sector of sanitary and environmental engineering.

D I P L O M A

For specialised PhD training

The Netherlands Research School for the
Socio-Economic and Natural Sciences of the Environment
(SENSE) declares that

Maribel Zapater Pereyra

born on 26 January 1984 in Piura, Peru

has successfully fulfilled all requirements of the
Educational Programme of SENSE.

Delft, 30 October 2015

the Chairman of the SENSE board

Prof. dr. Huub Rijnaarts

the SENSE Director of Education

Dr. Ad van Dommelen

KONINKLIJKE NEDERLANDSE
AKADEMIE VAN WETENSCHAPPEN

The SENSE Research School declares that Ms Maribel Zapater Pereyra has successfully fulfilled all requirements of the Educational PhD Programme of SENSE with a work load of 56.5 EC, including the following activities:

<u>SENSE PhD Courses</u>

○ Environmental Research in Context (2012)

○ Research in Context Activity: 'Co-organising International Conference: AGUA 2011, Ecosystems and Society, Universidad del Valle - Cali, Colombia)' (2011)

<u>Other PhD and Advanced MSc Courses</u>

○ Environment and health, Universidad del Valle, Colombia (2010)

○ Environmental processes fundamentals, Universidad del Valle, Colombia (2010)

○ Nanotechnology for water and wastewater treatment, UNESCO-IHE, The Netherlands (2014)

<u>Management and Didactic Skills Training</u>

○ Supervision of six MSc students, UNESCO-IHE, Delft (2012-2014)

○ Teaching assistant for the MSc course 'Eutrophication', UNESCO-IHE, Delft (2012)

○ Member of the International advisory committee of 5th International Symposium on Wetland Pollutant Dynamics and Control, Nantes, France (2013)

<u>Selection of Oral Presentations</u>

○ *Nanoparticle-beads, mussel shells and oyster shells. Potential materials for phosphorus removal in constructed wetlands.* 13th International Conference on Wetland Systems for Water Pollution Control, 25-29 November 2012, Perth, Australia

○ *Can a 9-cm depth horizontal flow constructed wetland, or "constructed wetroof", treat domestic wastewater?* 3rd IWA BeNeLux Young Water Professional Regional Conference, 2-4 October 2013, Esch-sur-Alzette, Luxembourg

○ *Reducing the constructed wetland (CW) footprint by means of a Duplex-CW.* 5th International Symposium on Wetland Pollutant Dynamics and Control, 13-17 October 2013, Nantes, France

○ *Nitrification and denitrification in a Duplex- constructed wetland.* 14th International Conference on Wetland Systems for Water Pollution Control, China, Meeting or Workshop, 12-16 October 2014, Shanghai, China

SENSE Coordinator PhD Education

Dr. ing. Monique Gulickx

T - #0400 - 101024 - C26 - 244/170/13 - PB - 9781138029309 - Gloss Lamination